THE

NEW AMERICAN ORDER

OF

ARITHMETIC:

CONTAINING THE

PRIMORDIAL SERIES OF INCREASE AND DECREASE;

WITH AN

AUTHENTIC, PLAIN, AND COMPREHENSIVE EXPOSITION

OF THE

Synthetical Formation and Relations of Numbers,

VARIOUSLY EXEMPLIFIED, IN ADAPTATION TO THE PURPOSES OF

ARTS AND COMMERCE

IN THE UNITED STATES.

BY D. McCURDY,

Late a Clerk in the U. S. Pension Office, Author of an Edition of Euclid's Elements, Chart of Geometry, First Lessons in Geometry, National Arithmetic, Columbian Tutor's Assistant, &c., &c.

Baltimore:

PUBLISHED BY ARMSTRONG & BERRY.

1850.

☞ *See Explanatory Notes and a Table of Contents at the end of the book.*

Stereotyped by C. Davison & Co.,
33 Gold street, N. Y.

A DIALOGUE

BETWEEN THE BOOK AND THE READER.

Book.—Why dost thou examine me so intensely, and turn me over so rudely? Thy duty is to handle thy book with care.

Reader.—I intended no disrespect; but it seems to me that a book very like you, in some parts, fell into my hands before to-day.

B.—Very probably. My first edition was published in Washington, in 1819; my second in Baltimore in 1826; but my author neglected both the former editions. Me he has renewed so much, that he thinks it sheer injustice not to give me a new name.

R.—But what was your author's reason for publishing your former editions, and now renewing you so essentially?

B.—He had taught in Pennsylvania, New Jersey, Maryland, and in the District of Columbia, from 1810; and found Jess's book the popular one of the time. It was to correct doctrinal errors in that book that he first prepared a system. The same errors were afterward copied by Stephen Pike, and followed by Bennett, Smiley, Lewis, Rose, and other respectable authorities; and they still exist.

R.—And what are those errors to which you allude?

B.—The principal error respects the doctrine of Compound Interest and Annuities, which should be based upon the involution of ratios; making the ratio the amount of a unit for the term of payment. The error, which was small in 1819, on account of the comparatively small business in ground-rents and stocks at that time, has greatly increased in importance. It consists in setting forth a false ratio for terms of payment less than a year. Those authors tell us that the second root of the annual ratio will be the ratio in half-yearly payments, the fourth root in quarter-yearly, &c. Now, either the second power of the second root, or the fourth power of the fourth root, if there should be no loss in extracting it, will restore the original number. So that their rule imposes vain labor; since it makes quarter and half-yearly the same as annual payments, and even worse; for the numbers being surds will lose a fraction in extracting their roots. The true ratio is found without extracting any root; and it is evident that instalments, rents, or periodical payments resulting from stocks of any kind, are of more advantage to the receiver by short payments than by long ones. It is also certain that most payments of this kind are made half, or quarter-yearly, or monthly.

R.—The statement you make is very plausible; but our books here in the North do not seem to dwell much upon the application of progressions to any purpose of business: will you lead me into the best way of understanding these progressions?

B.—Read the numbers of my pages in order; they form the prime series to some extent; the laws of this series contain the doctrines of arithmetic.

R.—I will follow your counsel. But why are not all these equal pages marked with the same number?

B.—The reason is, that number is the result of order; and the successive names express the distance of each from the first point of the series.

R.—The prime series, then, is a succession of names applied to a line of equal units; each name, or term, expressing in itself all the preceding units?

B.—It is even so; for of what form soever the units may be, arithmetic disposes them into lines or strata, and names them in order, beginning at some point, line, or surface.

R.—You seem to make arithmetic a system of series or progressions, which I have supposed to belong rather to algebra.

B.—Thy mistake is a very common one; the series are all numerical; every result of increase or decrease must be pursued along the lines of multiples or parts of a unit; the relations of numbers must become known before the symbols of algebra can be used to express them.

R.—What am I to consider proper views of proportions?

B.—Proportions are measured distances upon lines of numbers of the same species; in which the first term, or a part of it, measures the second as often as the third, or the same part of it, measures the fourth. Ratio is the number of times which the measure is applied.

R.—What is to be understood by the inductive system?

B.—Induction should lead by examples into the laws of the several series; passing from one series to another, when they are of different species or powers, furnished with the proper measure, reduced if necessary, to reach at once the required term.

But the Pestalozzian writers make induction a kind of *diluted proportion;* they apply their measure from the first given term to the unit; thence to the second given term and back to the unit; thence to the third and back to the unit; thence to the fourth and back; and so on to the last. Thus they use a double number of ratios, deal in larger measures, are often unnecessarily involved in fractions, and with much simplicity congratulate the system upon its relief from the prolixity of ratios and proportions.

R.—Is not that true induction which dispenses with mechanical rules and forms?

B.—True induction into chaos! The world's business, and its pleasures too, are a connected system of those forms which are reproachfully called mechanical. The measures of wealth and labor, or whatever goes under the name of property, are established by usage or enactment; in the comparison of these measures arithmetic should be practised. Hence will arise uniform movements or mechanical rules; and the Pestalozzian books which reject them are not guides to the regions of light, but mere devices to lead off the exuberant vitality of crowded schools, with the slightest possible explosion.

R.—I have heard many objections made to the use of words in elementary books, which are beyond the comprehension of children. Is it not possible to express in the plain English any sense that is to be conveyed?

B.—The plain English which embraces the arts and sciences embraces all languages, ancient and modern. But it will be very hard to find a word or phrase in any book, the meaning of which is not given by Webster or some of the lexicographers; and to exclude the necessary terms of a science is to break off the handle by which it is held, and transmitted from one to another.

R.—Is it not better to exercise the mind with early discussions than to store it to repletion with simple facts?

B.—Dost thou mean to discuss the mind or the facts? The ox ruminates after he has cropped the herbage. Knowledge increases by observation, experiment, and reflection; to each of these there is a time, and the portions of time are successive. The elements of science are simple sayings expressed in form; the mind which stores them largely, in due season, may use the subsequent times to dispose them in order, and comprehend their use; they are best known from their relations to each other, which are discovered after the mind has possession of them in sufficient number.

R.—What are the peculiar advantages of the mental arithmetic?

B.—Mental arithmetic (so called)is that which is the least mental; it suggests every operation, however minute; it amplifies all that mercantile experience had condensed into forms; and strews about the elements of arithmetic like the dilapidated architecture of a ruined temple. To thee these elements will be mental when they are restored to order, and the forms and process delineated upon thy memory, as the sun-painted image is portrayed upon the polished plate of the artist.

R.—To which of the nations does arithmetic belong?

B.—Spurious patriotism recommends books among us, on the plea that their examples are given exclusively in Federal money; and this may be the bearing of thy question. But arithmetic is based upon duration, extension, and gravitation, which are immunities of the Supreme System. Every country requires certain measures of the labor and wealth of its inhabitants, and being under no restraint, they severally adopt such numbers as they please; and these numbers form their arithmetic. Commercial nations ought to learn each other's systems, especially in regard to money and the measures of such commodities as they exchange with each other.

R.—I perceive at the head of your pages a line of numbers beginning at 1 and increasing by 1; also below each number a fraction having for its numerator 1, and for its denominator the same number; will you state the uses of these two lines?

B.—The lines in question are the multiples and parts of a unit which are the reciprocals of each other. To state their uses would be to solve the science of arithmetic in all its parts; which is possible alone from this arrangement. The want of this illustration has ever been a defect in all systems of arithmetic.

R.—You have furnished a track from which it is hard to deviate; and your directions are sufficiently plain; any one who can read mile-stones may do all that you prescribe.

B.—The simplest process of reasoning is counting; children perform it as soon as they learn to speak. It is also the most certain method and the highest; for Solomon, with all his wisdom, resorted to it. He "counted one by one to find out the matter." Moreover, counting implies being, arrangement, motion, and extension, and puts to flight the delusive notion that the truths of arithmetic and geometry would exist if there were no creation—nothing to be measured or counted. For what is truth but the appropriate relation of one *thing* to another; that which leaves nothing defective, redundant, or disproportioned in the great scale of *beings*. There is no truth where there is nothing to be made true.

1	2	3	4	5	6	7	8	9	10
$\frac{1}{1}$	$\frac{1}{2}$	$\frac{1}{3}$	$\frac{1}{4}$	$\frac{1}{5}$	$\frac{1}{6}$	$\frac{1}{7}$	$\frac{1}{8}$	$\frac{1}{9}$	$\frac{1}{10}$

INTRODUCTION.

SPACE absolute, or vacuity, is not a subject of arrangement or law, it has no points of beginning or end, and cannot be measured or numbered.

But land, or water, air, light, heat, sound, time, the planets, comets, occupying or passing through space, are subjects of measure or number. Time, however, is not a substantive quantity, and it must therefore cease with the motion of bodies by which it is measured.

Every atom, and every organized body in the universe, counts *one*, a *unit;* a grain of sand, a rose, a house, the state, a star, the universe itself is *one*.

Every unit occupies a space from which it excludes all other units; therefore order and extension are essential to numbers.

Independent units are not properly called numbers; those units only which form the terms of a series are entitled to this appellation. Every unit in a series is similar to the first.

Arithmetic classifies all things, according to their species, magnitude, and other similarities; then places them in lines, calling them *one*, *two*, *three*, *four*, *five*, *six*, &c.

These names, or terms, are numbers. Every number declares its own distance from the first point of the series, in measures or lengths of the first unit.

The *first* or *prime* series is that whose first term is *one*, and constant increase *one*.

The prime series contains all numbers, because a unit is its lowest term, and a unit is the lowest number.

A fraction is not a number, because it is not excluded

11	12	13	14	15	16	17	18	19	20
$\frac{1}{11}$	$\frac{1}{12}$	$\frac{1}{13}$	$\frac{1}{14}$	$\frac{1}{15}$	$\frac{1}{16}$	$\frac{1}{17}$	$\frac{1}{18}$	$\frac{1}{19}$	$\frac{1}{20}$

from the place occupied by the number of which it is a part; but if the parts be plural, and one of them be made the first term of a series, the fraction becomes a number.

Any number not compared with another is *one;* a fraction also is one when measured by itself.

Induction, in arithmetic, is that form of practice which leads into the laws of the several series.

Of the Formation of Series.

All things seen or thought of in the universe, belong to the department of arithmetic; but some book-makers claim patronage for their books, because they contain only a domestic selection. We will therefore, to be even with these patriots, confine our examples chiefly to the home circle, and treat of things which come to market and bring the dollars.

But first, where are our numbers?

The *free* atoms or *things* are not numbers, and to catch them, and place them in lines or series, is another business.

Well, we have found an expedient. We can inclose a space with black lines at the head of our pages; we can divide this space, first into halves by a line drawn through the middle of it, and then each half into ten equal parts.

These little spots of paper will represent things in general, and we may call them dollars, acres, bushels, gallons, yards, pounds, miles, boys, toys, tops, or whatever else we please.

All numbers are represented by the ten digits, viz.: nought, 0; one, 1; two, 2; three, 3; four, 4; five, 5; six, 6; seven, 7; eight, 8; nine, 9; and their combinations; and they are placed in order upon the upper headline of these pages.

Of the Use to be made of the Prime Series.

To count the spaces, or, which is the same thing, to repeat the terms of the prime series, is the young student's first lesson in arithmetic.

21	22	23	24	25	26	27	28	29	30
$\frac{1}{21}$	$\frac{1}{22}$	$\frac{1}{23}$	$\frac{1}{24}$	$\frac{1}{25}$	$\frac{1}{26}$	$\frac{1}{27}$	$\frac{1}{28}$	$\frac{1}{29}$	$\frac{1}{30}$

Case 1. When the first term and common increase are the same.

1. The class begins at 1 on the line of numbers; saying, one, two, three, four, five, six, seven, &c., in succession, not in chorus, and thus proceed in order to the end of the book.

It would be tedious to say 1 and 1 are 2; 2 and 1 are 3, &c.; but this addition is actually done by repeating the terms in successive order.

2. The class will now proceed to form the series of even numbers, beginning at 2 and adding 2s; thus, the first says 2; the second, 2 and 2 are 4; the third, 4 and 2 are 6, &c., to the end of the book.

3. After the same manner is formed the series of 3s: thus, 3 and 3 are 6; 6 and 3 are 9; 9 and 3 are 12, &c.

4. Begin now at 4, and add 4s.
5. Begin at 5, and add 5s.
6. Begin at 6, and add 6s.
7. Begin at 7, and add 7s.
8. Begin at 8, and add 8s.
9. Begin at 9, and add 9s.
10. Begin at 10, and add 10s.
11. Begin at 11, and add 11s.
12. Begin at 12, and add 12s.

The class study the preceding exercises with open book, but the recitations should be made with book closed.

Let the additions be carried out, in every example, to the end of the book.

Case 2. When the first term and common increase are different.

1. Let 1 be the first term, and let the class proceed to form the series of odd numbers by adding 2s; thus, the first says *one;* the second, 1 and 2 are 3; the third, 3 and 2 are 5; the fourth, 5 and 2 are 7, &c.

2. In like manner, beginning with 1, the class adds 3s; thus, 1; 1 and 3 are 4; 4 and 3 are 7; 7 and 3 are 10, &c. Carry these additions along the line of numbers to the end.

31	32	33	34	35	36	37	38	39	40
$\frac{1}{31}$	$\frac{1}{32}$	$\frac{1}{33}$	$\frac{1}{34}$	$\frac{1}{35}$	$\frac{1}{36}$	$\frac{1}{37}$	$\frac{1}{38}$	$\frac{1}{39}$	$\frac{1}{40}$

3. To 1 add 4s.
4. To 1 add 5s.
5. To 1 add 6s.
6. To 1 add 7s.
7. To 1 add 8s.
8. To 1 add 9s.
9. To 1 add 10s.
10. To 1 add 11s.

The preceding exercises may be reversed by subtraction; thus, begin at the highest term, which may be found upon the last page, and measure back upon the line of numbers with the common increase, which in subtraction is the common difference, to the first point of the series.

Case 3. To form series with a variable increase.

1. The prime series is first in order for this purpose. In class, the first says 1; the second, 1 and 2 are 3; the third, 3 and 3 are 6; the fourth, 6 and 4 are 10, &c.

The terms of this last series are triangular numbers; viz., 1, 3, 6, 10, 15, 21, &c., which may be shown with pencil points, or grains of corn, on a slate. Carry the addition up to thirty terms, upon the line of numbers.

2. Having set down thirty terms of the last series on a slate, the class will now proceed to form a series of pyramidal numbers by adding the terms of the triangular series together, supposing the units to be balls; thus, 1: 1 and 3 are 4; 4 and 6 are 10; 10 and 10 are 20; 20 and 15 are 35, &c.

These numbers are used in the ternary combination lotteries: they form a triangular pyramid.

3. The terms of the series of odd numbers are next to be used in forming a series with variable increase; these are, 1, 3, 5, 7, 9, 11, &c.

In class, the first says 1; the second, 1 and 3 are 4; the third, 4 and 5 are 9; the fourth, 9 and 7 are 16, &c., up to thirty terms.

The terms of the series now formed are squares; viz., 1, 4, 9, 16, 25, &c., and supposing the units of each term to be solids, as balls, the courses may be laid upon each other so as to form a square pyramid.

4. Having set down about thirty terms of the series of square numbers, the class will proceed to form a series

41	42	43	44	45	46	47	48	49	50
$\frac{1}{41}$	$\frac{1}{42}$	$\frac{1}{43}$	$\frac{1}{44}$	$\frac{1}{45}$	$\frac{1}{46}$	$\frac{1}{47}$	$\frac{1}{48}$	$\frac{1}{49}$	$\frac{1}{50}$

of pyramidal numbers; thus, the first says 1; the second, 1 and 4 are 5; the third, 5 and 9 are 14; the fourth, 14 and 16 are 30, &c.

5. Let us now turn to the series of 6s (Case 1), the terms of which are 6, 12, 18, 24, 30, &c. With these terms the class will form a new series; thus, the first says 1; the second, 1 and 6 are 7; the third, 7 and 12 are 19; the fourth, 19 and 18 are 37; the fifth, 37 and 24 are 61, &c., up to twenty terms.

The terms 7, 19, 37, 61, 91, &c., now formed, are the differences of the terms of the series of cubes; and after setting down as many of them as may be convenient, the class proceeds to form the series of cubes; thus, the first says 1; the second, 1 and 7 are 8; the third, 8 and 19 are 27; the fourth, 27 and 37 are 64; the fifth, 64 and 61 are 125, &c.

The terms 1, 8, 27, 64, 125, 216, &c., are the cubes of the terms of the prime series, 1, 2, 3, 4, 5, 6, &c.

Of Numeral Letters.

The letters I, V, X, L, C, D, M, are used, after the manner of the ancient Romans, to denote numbers; as follows, viz.:

I, one; V, five; X, ten; L, fifty; C, one hundred; D, five hundred; M, one thousand. These, by combination, form other numbers; thus, to increase a greater letter annex the less; as VI, six; XV, fifteen; LX, sixty; CX, one hundred and ten. To lessen the value of the greater prefix the less; as IV, four; IX, nine; XL, forty; XC, ninety, &c.

The replication of the same letter repeats the quantity; thus, CCC is three hundred, XX is twenty, II is two.

MDCCCXLVII denotes 1847. This was evidently a system which required improvement.

Of the Arabic Figures.

To the Moors both Europe and America are indebted for the use of numerals less prolix than either the Grecian or Roman notation. They are called digiti (fingers)

51	52	53	54	55	56	57	58	59	60
$\frac{1}{51}$	$\frac{1}{52}$	$\frac{1}{53}$	$\frac{1}{54}$	$\frac{1}{55}$	$\frac{1}{56}$	$\frac{1}{57}$	$\frac{1}{58}$	$\frac{1}{59}$	$\frac{1}{60}$

because there are ten of them. Their names are already mentioned.

Of the Composition of High Numbers.

High numbers are composed by placing numbers of different ranks in the same line; the line is then divided into periods of six, and half periods of three places.

Different ranks are those which vary in distance from the units' place: viz., 1, 10, 100, 1000, 10000, &c.

The decimal notation reverses the order of these numbers, and takes away the 0s; which are only counters of the places to unity; thus,

+6 +5 +4 +3 +2 +1 +0
1000000, 100000, 10000, 1000, 100, 10, 1.

These are the decimal numbers reversed; and the series, 6, 5, 4, 3, 2, 1, 0, placed over the significant figures, indicate how far each is removed from the units' place.

Now taking away the 0s, and joining the 1s, thus, 1.111,111, it is plain that the 1s perform for each other the office of counting the ranks; and all the numbers above combined will make the single term one million, one hundred and eleven thousand, one hundred and eleven.

The periods are called units, millions, billions, trillions, quadrillions, quintillions, sextillions, septillions, &c.; namely, units the first period, and millions of the first, second, third, fourth, fifth, sixth, &c., order, for the rest.

The French division of periods is not supported by the etymology of the terms, in our language.

The several ranks and periods are arranged in their proper order, and named in the annexed

TABLE.

Read h hundred, t tens, u units. But in reading, the term unit is usually omitted, because it is the denominator of every number, and always understood.

trillions. thousands. billions. thousands. millions. thousands. units.
u h t u h t u h t u h t u h t u h t u
9.876,543.210,987.654,321.

61	62	63	64	65	66	67	68	69	70
$\frac{1}{61}$	$\frac{1}{62}$	$\frac{1}{63}$	$\frac{1}{64}$	$\frac{1}{65}$	$\frac{1}{66}$	$\frac{1}{67}$	$\frac{1}{68}$	$\frac{1}{69}$	$\frac{1}{70}$

This number, which is too large for any human purpose, reads thus: 9 trillions, 876 thousand 543 billions, 210 thousand 987 millions, 654 thousand 321.

In reversing the numbers, the units' rank is brought to the right of the line, and the periods follow the same rule.

In class: Ten 1s are ten, ten 10s are 100, ten 100s are 1000. Also, one thousand 1000s are 1000000, one million 1000000s are 1 billion, one million of billions are 1 trillion, &c.

Write in figures, using 0s to count the ranks, the following numbers: viz., one, twenty, three hundred, four thousand, fifty thousand, six hundred thousand, seven millions, eighty millions, nine hundred millions, ten thousand millions, two hundred thousand millions, three billions, forty billions, five hundred billions, six thousand billions, seventy thousand billions, eight hundred thousand billions, nine trillions.

What is the first rank in every half period called?—the second?—the third? What is the first half of each period called?—the second? What is the first period called?—the second?—the third?—the fourth?—the fifth? &c.

The Operative Signs herein used are the following:

+ The erect cross.
× The oblique cross.
— The dash.
÷ The dash and points.
: The colon.
: : The double colon.
= The parallels.
. The units' point.
√ The root.
—— The vinculum.

Operations are indicated by these signs, as follows: viz.,
Addition; thus, 8+6 signifies 8 and 6, or 8 plus 6.
Multiplication; thus, 7×2 denotes 7 times 2.
Subtraction; thus, 8—6 denotes 8 less 6.
Division; thus, 14÷2 denotes 14 divided by 2.
Evolution; thus, √16, denotes the square root of 16.
Ratio; thus, 2 : 4, or 2 is to 4.
Proportion; thus, 2 : 4 : : 3 : 6, or 2 is to 4 as 3 is to 6.

71	72	73	74	75	76	77	78	79	80
$\frac{1}{71}$	$\frac{1}{72}$	$\frac{1}{73}$	$\frac{1}{74}$	$\frac{1}{75}$	$\frac{1}{76}$	$\frac{1}{77}$	$\frac{1}{78}$	$\frac{1}{79}$	$\frac{1}{80}$

Equality; thus, $8+6=14$, or 8 and 6 are 14; also, $7\times2=14$, or 7 times 2 are 14; and $8-6=2$, or 8 less 6 equals 2; and $14\div2=7$, or 14 divided by 2 equals 7; and $\sqrt{16}=4$, or the square root of 16 is equal to 4. The units' point marks the units' place when there are decimal fractions in the quantity; thus, 1.06 reads 1 and 6 hundredths. The vinculum joins two or more quantities in one quantity; thus, $\overline{1.06\times.06}+1.06$, or $\overline{p\times r}+p$.

Fuller explanations will be found where the signs are used.

Theorems showing the properties of the digit 9.

1. Any number, the sum of whose digits is a multiple of 9 is itself a multiple of 9.

The digits of 432, viz., 4, 3, and 2, are 9, that is, the first multiple of 9; but $432=400+30+2=\overline{4\times99}+\overline{3\times9}+\overline{4+3+2}$; and these together make the 48th multiple of 9.

2. Any number, the sum of whose digits exceeds a multiple of 9, does itself equally exceed a multiple of 9.

The digits of 226 are 2, 2, 6; and $2+2+6=10=9+1$; but $226=200+20+6=\overline{2\times99}+\overline{2\times9}+\overline{2+2+6}$; and these together make the 25th multiple of 9 with an excess of 1.

Wherefore every number exceeding 9 may be expressed by 9s and an excess less than 9. But this property is common to all numbers; for every number may be expressed by 10s, or 10s and an excess less than 10; and by 8s, or 8s and an excess less than 8. The demonstration of this is obvious by counting off the terms of the prime series by 1s, 2s, 3s, 4s, &c., and by setting down the measuring number for every time it is repeated. The formation of series is therefore the simplest illustration of the composition of numbers.

81	82	83	84	85	86	87	88	89	90
$\frac{1}{81}$	$\frac{1}{82}$	$\frac{1}{83}$	$\frac{1}{84}$	$\frac{1}{85}$	$\frac{1}{86}$	$\frac{1}{87}$	$\frac{1}{88}$	$\frac{1}{89}$	$\frac{1}{90}$

SIMPLE ADDITION.

To add is to find a term in the line of numbers, or the line of the prime series, to which all the units of the given numbers will extend.

Thus 6 and 7 extend to 13, 9 and 8 reach to 17.

In class: How far upon the line of numbers will the units of the following numbers extend; viz.,

1, 2, and 3?	9, 10, and 11?	17, 18, and 19?
2, 3, and 4?	10, 11, and 12?	18, 19, and 20?
3, 4, and 5?	11, 12, and 13?	19, 20, and 21?
4, 5, and 6?	12, 13, and 14?	20, 21, and 22?
5, 6, and 7?	13, 14, and 15?	21, 22, and 23?
6, 7, and 8?	14, 15, and 16?	22, 23, and 24?
7, 8, and 9?	15, 16, and 17?	23, 24, and 25?
8, 9, and 10?	16, 17, and 18?	24, 25, and 26?

Here we have arrived at numbers too high for oral answers; but in case of inability, the reader may resort to the line of numbers for assistance.

It now becomes necessary to recur to the principles of the decimal notation: viz., ten 1s are 10; ten 10s are 100; ten 100s are 1000, &c. And since 9, the highest digit, supplies every rank with its full complement of units, if another number be added to 9 the rank overflows, and ten of the units, or twenty, or thirty, as the case may be, are carried to the next rank on the left, and all the units over what completes one, two, three, or more tens, are set down in the rank which has been added; thus,

t u	h t u	u h t u	t u h t u	h t u h t u
9	99	999	9999	99999
1	2	3	4	5
10	101	1002	10003	100004

The number carried is one for every ten.

The same overflow of units, from rank to rank, occurs from addition when the ranks are filled by the sums of several smaller digits instead of 9.

91	92	93	94	95	96	97	98	99	100
$\frac{1}{91}$	$\frac{1}{92}$	$\frac{1}{93}$	$\frac{1}{94}$	$\frac{1}{95}$	$\frac{1}{96}$	$\frac{1}{97}$	$\frac{1}{98}$	$\frac{1}{99}$	$\frac{1}{100}$

t u	h t u	u h t u	t u h t u	h t u h t u
1	12	123	1234	12345
2	34	456	2345	23456
3	56	789	3456	34567
4	78	012	4567	45678
10	180			

From the foregoing remarks and examples is deduced the following

Rule.

Place units under units, tens under tens, &c.; begin at the right hand and add upward; set down the sum of each column, if less than ten; but if the sum consist of two figures or more, set down only the units, and carry the rest to the next column, setting down all at the last.

Proof. Add the columns downward, or divide the work into parts; take the sum of each part, and the sum of these sums shall equal the whole amount.

Otherwise: The excess of 9s in the sum of the digits of the given numbers will equal the same excess in the whole amount.

Examples for the Slate.

$	cts.	mills	yds.	£
5	25	125	45	3368
7	50	375	364	1296
9	75	625	276	6250
11	100	875	216	926
13	125	1000	829	7854
15	150	1225	93	2240
?	?	?	?	?

Note. The answers of some of the following questions are given in Subtraction.

6. Statements from the Custom House, New York, showing the amount in dollars of importations for September, 1846 and 1847.

1846		1847	
$600,849	Free goods.	$916,109	Free goods.
5.272,923	Dutiable.	8.111,845	Dutiable.
10,014	Specie.	94,546	Specie.
$	? in 1846.	$	? in 1847.

101	102	103	104	105	106	107	108	109	110
$\frac{1}{101}$	$\frac{1}{102}$	$\frac{1}{103}$	$\frac{1}{104}$	$\frac{1}{105}$	$\frac{1}{106}$	$\frac{1}{107}$	$\frac{1}{108}$	$\frac{1}{109}$	$\frac{1}{110}$

7. Duties paid at the Custom House, New York, during three months in 1846 and 1847.

\$14.611,525	July and August.	\$16.182,251
1.542,455	September.	2.096,604
\$?	in 1846.	\$? in 1847.

8. Exports from New York in September, 1846 and 1847.

\$2.238,401	Domestic.	\$2.672,452
388,169	Foreign.	193,375
2,255	Specie.	350,925
\$?	in 1846.	\$? in 1847.

9. Imports to New York for ten months, commencing December 1, 1845 and 1846.

\$4.076,672	December,	\$4.878,655
5.283,010	January,	6.068,999
4.749,091	February,	7.409,637
9.812,491	March,	8.177,141
6.440,815	April,	13.723,526
5.488,397	May,	7.033,713
5.873,655	June,	6.638,280
6.195,709	July,	9.106,399
8.457,124	August,	13.574,041
5.833,815	September,	2.122,500
\$?	in 1845.	\$? in 1846.

10. Europe measures 3387109 square miles; Asia, 16728002; Africa, 11652442; America, 16504204; Australia and the rest, 4164420; how many square miles of land on our globe? *Ans.* 52436177 sq. m.

11. Merchants use ruled books to keep straight their money columns, for addition.

Dols.	cts.		Dols.	cts.		Dols.	cts.
3456	75		3957	28		2438	63
9632	28		8275	93		4823	36
7459	64		2857	39		9834	29
5873	29		5728	75		3976	87
8934	22		7682	48		2569	45
	?			?			?

111	112	113	114	115	116	117	118	119	120
$\frac{1}{111}$	$\frac{1}{112}$	$\frac{1}{113}$	$\frac{1}{114}$	$\frac{1}{115}$	$\frac{1}{116}$	$\frac{1}{117}$	$\frac{1}{118}$	$\frac{1}{119}$	$\frac{1}{120}$

12. Exports from New York for ten months, commencing December 1, 1845 and 1846.

1845	Month	1846
$2.796,314	December,	$4.065,821
2.122,696	January,	3.192,626
1.972,545	February,	3.468,000
1.309,598	March,	4.146,896
2.828,880	April,	3.933,674
3.114,549	May,	4.159,864
4.062,249	June,	7.256,290
3.119,295	July,	6.837,340
2.678,627	August,	5.045,108
2.628,825	September,	3.216,752
$? in 1845.		$? in 1846.

13. Questions in chronology, showing the order of succession and reign of English dynasties and kings, commencing with the old Saxon, A. D., 827.

OLD SAXON, 827
Egbert's, 11
Ethelwolf's, 19
Ethelbald's, 3
Ethelbert's, 6
Ethelred's, 6
Alfred's, 28
Edward's, 25
Athelstan's, 16
Edmund's, 7
Edred's, 7
Edwin's, 4
Edgar's, 16
Edward's, 4
Ethelred's, 37
Edmund's, 1

DANISH, ?
Canute's, 19
Harold's, 3
Hardicanute's, 3

NEW SAXON, ?
Edward's, 24
Harold's, 00

NORMAN, ?
William's, 21
William's, 13
Henry's, 35
Stephen's, 19

PLANTAGENET, ?
Henry's, 35
Richard's, 10
John's, 17
Henry's, 56
Edward's, 35
Edward's, 20
Edward's, 50
Richard's, 22

LANCASTER, ?
Henry's, 14
Henry's, 9
Henry's, 39

YORKERS, ?
Edward's, 22
Edward's, 00
Richard's, 3

TUDORS, ?
Henry's, 24
Henry's, 38
Edward's, 6
Mary's, 5
Elizabeth's, 45

STUARTS, ?
James', 22
Charles', 24
(Cromwell's,) 9
Charles', 25
James', 4

REVOLUTION, ?
William and Mary's, 14
Anne's, 12

HANOVER, ?
George's, 13
George's, 33
George's, 60
George's, 10
William's, 7
Victoria's, 10

In class: In what year did the Old Saxon dynasty commence?—the Danish?—the New Saxon?—the Norman?—the Plantagenet?—the Lancaster?—the York?—the Tudor?—the Stuart?—the Hanoverian? What year is the 10th of Victoria?

14. Succession and reign of the sovereigns of France, commencing A. D., 768, at the accession of Charlemagne.

121	122	123	124	125	126	127	128	129	130
$\frac{1}{121}$	$\frac{1}{122}$	$\frac{1}{123}$	$\frac{1}{124}$	$\frac{1}{125}$	$\frac{1}{126}$	$\frac{1}{127}$	$\frac{1}{128}$	$\frac{1}{129}$	$\frac{1}{130}$

CARLOVINGIAN,	768
Charlemagne,	46
Louis',	26
Charles',	37
Louis',	2
Louis',	3
Carloman's,	2
Charles',	4
Eudes',	10
Charles',	23
Robert's,	2
Rodolph's,	13
Louis',	18
Lothaire's,	32
Louis',	2
	——

CAPETIAN,	?
Hugh Capet's,	8
Robert's,	35
Henry's,	29
Philip's,	48
Louis',	29
Louis',	43
Philip's,	43
Louis',	3
Louis',	44
Philip's,	15
Philip's,	29
Louis,'	2
John's,	0
Philip's,	5
Charles',	7
	——

VALOISE,	?
Philip's,	22
John's,	14
Charles',	16
Charles',	42
Charles',	39
Louis,'	22
Charles,'	15
Louis',	17
ANGOULEME,	?
Francis',	32
Henry's,	12
Francis',	1
Charles',	14
Henry's,	15
BOURBON,	?

Henry's,	21
Louis',	33
Louis',	72
Louis',	60
Louis',	18
REPUBLIC,	?
Directory and Consulate,	11
EMPIRE,	?
Napoleon's	10
Louis',	10
Charles',	6
REVOLUTION,	?
Louis Philip's,	17

In class: In what year did the Carlovingian dynasty commence in France?—the Capetian?—the house of Valoise?—of Angouleme?—of Bourbon?—the Republic?—the Empire?—the Revolution?—What year of our Lord is the 17th year of Louis Philip?

15. The census of the United States, taken in 1840, makes the population of the several States as follows, viz.:

Maine had	501,793	Maryland,	469,232	Ohio,	1.519,467
N. Hamp.,	284,574	Virginia,	1.239,797	Indiana,	685,866
Mass.,	737,699	N. Carolina,	753,419	Illinois,	476,183
Vermont,	291,948	S. Carolina,	594,398	Missouri,	383,702
R. Island,	108,830	Georgia,	691,392	Arkansas,	97,574
Connecticut,	309 978	Alabama,	590,756	Michigan,	212,267
New York,	2.428 921	Mississippi,	375,651	Florida,	54,477
N. Jersey,	373,306	Louisiana,	352,411	Wisconsin T.	30,945
Penn.,	1.724,033	Tennessee,	829,210	Iowa T.,	43,112
Delaware,	78,085	Kentucky,	729,828	Dis. Col.,	43,712
	?		?		?

In class: How many States and Territories are mentioned above? What was the population of the first ten?—of the second ten?—of the third ten?—of the whole number? What is the meaning of taking the census?

131	132	133	134	135	136	137	138	139	140
$\frac{1}{131}$	$\frac{1}{132}$	$\frac{1}{133}$	$\frac{1}{134}$	$\frac{1}{135}$	$\frac{1}{136}$	$\frac{1}{137}$	$\frac{1}{138}$	$\frac{1}{139}$	$\frac{1}{140}$

SIMPLE SUBTRACTION.

To subtract is to take a less number from a greater; or, to find how far the units of one number extend beyond the units of another number upon the line of the prime series, or line of numbers.

Hence, there are two terms given and one term sought.

1. The *minuend* is the greater given number
2. The *subtrahend* is the less given number.
3. The *difference*, *remainder*, or *residue*, is the number sought.

The terms are of the same kind; or, at least, they must be such as to come under some common name, and have the same measuring unit.

In class: 1. Begin at 1000 upon the line of numbers, and count off all the units, one by one, until none are left. This may be done by repeating the terms of the series backward; thus, 1000, 999, 998, 997, &c.

2. Begin at 1000 upon the line of numbers, and count off all the units, two by two, until none are left. This may be done by repeating the terms of the series of even numbers; thus, 1000, 998, 996, 994, 992, 990, &c.

After the same manner, other numbers are subtracted.

3. Count off 1000 by 3s.
4. Count off 1000 by 4s.
5. Count off 1000 by 5s.
6. Count off 1000 by 6s.
7. Count off 1000 by 7s.
8. Count off 1000 by 8s.
9. Count off 1000 by 9s.
10. Count off 1000 by 10s.
11. Count off 1000 by 11s.
12. Count off 1000 by 12s.

Here we arrive at numbers too high for oral responses, and which require a statement in accordance with the ranks of which they consist; as units, tens, hundreds, &c.

Rule.

Place the less given number under the greater; so that units come under units, tens under tens, &c. Begin at the units' place, taking the lower figure from the one

141	142	143	144	145	146	147	148	149	150
$\frac{1}{141}$	$\frac{1}{142}$	$\frac{1}{143}$	$\frac{1}{144}$	$\frac{1}{145}$	$\frac{1}{146}$	$\frac{1}{147}$	$\frac{1}{148}$	$\frac{1}{149}$	$\frac{1}{150}$

above it, if possible; if not, add ten to the upper and subtract; set down the remainder; add 1 to the next lower figure for the 10 (if any) added to the upper: continue the same process to the last rank on the left.

Proof 1. Remainder+Subtrahend=the Minuend.

2. Minuend−Remainder=Subtrahend.

3. The excess of 9s in the digits of the minuend is equal to the excess of 9s in the digits of the Subtrahend and Remainder.

Examples for the Slate.

	h t u h t u		h t u h t u		h t u h t u	
From	854632	Minu.	572943	Minu.	839154	=3
Take	231496	Subtra.	349275	Subtra.	451938	=3
Diff.	623136	Rem.	223668	Rem.	387216	
	854632	Proof 1.	349275	Proof 2.	Proof 3.	Excess.

4.	754628 287569	5.	895296 729658	6.	965827 519876
7.	1234567 765432	8.	6534210 3456789	9.	4932487 3497842

10. The value of goods imported into New York, in September, 1846, was 5883786 dollars; and in September, 1847, 9122500 dollars; what was the increase?
Ans. 3238714.

11. The duties paid at New York in July, August, and September, 1846, amounted to 16153980; and in the same months of 1847 the duties paid at that custom house amounted to 18278855 dollars; how much was the increase? *Ans.* 2124875 dols.

12. The exports from New York, in domestic and foreign merchandise and specie, in September, 1846, amounted to 2628825 dollars, and in September, 1847, to 3216752 dollars; how much was the increase?
Ans. 587927 dols.

151	152	153	154	155	156	157	158	159	160
$\frac{1}{151}$	$\frac{1}{152}$	$\frac{1}{153}$	$\frac{1}{154}$	$\frac{1}{155}$	$\frac{1}{156}$	$\frac{1}{157}$	$\frac{1}{158}$	$\frac{1}{159}$	$\frac{1}{160}$

13. The value of imports to New York, for ten months, commencing Dec. 1, 1845, was 62210779 dollars; and for ten months, commencing Dec. 1, 1846, was 78732891 dollars; how much was the increase?

Ans. 16522112 dols.

14. The value of exports from New York, for ten months, commencing Dec. 1, 1845, was 26633578 dollars; and for ten months, commencing Dec. 1, 1846, was 45322371 dollars; how much was the increase?

Ans. 18688793 dols.

15. By the census of 1820, the population of the United States amounted to 9625734; and by the census of 1840, to 17012566; what was the increase?

Ans. 7386832.

16. In 1815, the revenue of the U. S. was 37656436 dollars; in 1820, it was only 15284546; also of the former amount 14491739 dollars was received at New York; how much was received at all other ports? and how much the decrease? *Ans.* Other ports, 23164697 dollars; decrease, $22371890.

17. The whole revenue of the U. S. from 1789 to 1815, was 247019302 dollars; the expenditure during that time was 184719336 dollars; how many years intervened? and how much surplus left in the treasury?

Ans. 62299966 dols. and 26 years.

18. In 1816, the U. S. owed 123016375 dollars; and in 1824, only 90177962 dollars; how much of the debt was paid in the mean time? *Ans.* 32838413 dols.

19. Receipts and expenses of the Baltimore and Ohio Railroad and Washington Branch, for two years,

—ending Sept. 30, 1843.		—ending Sept. 30, 1844.	
Receipts,	$575,235 08	Receipts,	$658,619 98
Expenses,	287,153 72	Expenses,	294,833 29
	$?		$?

20. How much did the profits of 1844 exceed those of 1843, by the estimate in example 19? *Ans.* $75,705 33.

161	162	163	164	165	166	167	168	169	170
$\frac{1}{161}$	$\frac{1}{162}$	$\frac{1}{163}$	$\frac{1}{164}$	$\frac{1}{165}$	$\frac{1}{166}$	$\frac{1}{167}$	$\frac{1}{168}$	$\frac{1}{169}$	$\frac{1}{170}$

SIMPLE MULTIPLICATION.

To multiply is to apply a given number of units of the same kind, repeatedly, to the line of numbers, to find any required multiple of a given number.

Every number (waving etymology) is the first multiple of itself, its double is the second, its triple is the third, &c.

The *line of numbers* and *prime series* are the same; every number and every multiple of a number, is a term of this series.

The multiples of the first twelve numbers were found by adding, in Case 1, Introduction.

Every term of the prime series, that is, every number, is a multiple of 1. Every second term is a multiple of 2.

Every 3d is a multiple of 3.
Every 4th is a multiple of 4.
Every 5th is a multiple of 5.
Every 6th is a multiple of 6.
Every 7th is a multiple of 7.
Every 8th is a multiple of 8.
Every 9th is a multiple of 9.
Every 10th is a mult. of 10.
Every 11th is a mult. of 11.
Every 12th is a mult. of 12.

A common multiple is a term to which two or more terms extend.

Class Exercise upon the Line of Numbers.

Repeat 1260 multiples of 2.
Repeat 840 multiples of 3.
Repeat 630 multiples of 4.
Repeat 504 multiples of 5.
Repeat 420 multiples of 6.
Repeat 360 multiples of 7.
Repeat 315 multiples of 8.
Repeat 280 multiples of 9.
Repeat 252 multiples of 10.
Repeat 210 multiples of 12.

The last multiple of each of these series is the least common multiple of all the numbers above mentioned. What is it?

The exercise above prescribed is voluminous, but very easy; teachers may use their discretion as to the extent to which they will carry it in class.

171	172	173	174	175	176	177	178	179	180
$\frac{1}{171}$	$\frac{1}{172}$	$\frac{1}{173}$	$\frac{1}{174}$	$\frac{1}{175}$	$\frac{1}{176}$	$\frac{1}{177}$	$\frac{1}{178}$	$\frac{1}{179}$	$\frac{1}{180}$

In Multiplication there are two terms given: it is evident, also, that a unit is always given.

The *multiplicand* is the number to be applied (md.).

The *multiplier* is the number of applications (mr.).

The *product* is the required multiple (pt.).

Since md×mr=1×pt, therefore, 1 : md : : mr : pt.

The multiplicand and product are quantities of the same kind; the multiplier is also of the same kind in parcels, of which the multiplicand is one.

It is more convenient to use the less number as multiplier; and a change of the factors will not affect the product; for 12 boys may stand in 3 rows of 4 each, or 4 rows of 3 each; so that 4 times 3 equal 3 times 4.

The *factors* are the multiplicand and multiplier, and when the units of a product take the form of the table below, the product is called a rectangle.

Table showing 12 *multiples of the first* 12 *numbers.*

1	2	3	4	5	6	7	8	9	10	11	12
2	4	6	8	10	12	14	16	18	20	22	24
3	6	9	12	15	18	21	24	27	30	33	36
4	8	12	16	20	24	28	32	36	40	44	48
5	10	15	20	25	30	35	40	45	50	55	60
6	12	18	24	30	36	42	48	54	60	66	72
7	14	21	28	35	42	49	56	63	70	77	84
8	16	24	32	40	48	56	64	72	80	88	96
9	18	27	36	45	54	63	72	81	90	99	108
10	20	30	40	50	60	70	80	90	100	110	120
11	22	33	44	55	66	77	88	99	110	121	132
12	24	36	48	60	72	84	96	108	120	132	144

In this table the factors are the terms of the prime series at the top and on the left; the lines leading from the factors meet at the products, which consist of as many units as there are spaces cut off.

Every number, except the first, is a composite number;

181	182	183	184	185	186	187	188	189	190
$\frac{1}{181}$	$\frac{1}{182}$	$\frac{1}{183}$	$\frac{1}{184}$	$\frac{1}{185}$	$\frac{1}{186}$	$\frac{1}{187}$	$\frac{1}{188}$	$\frac{1}{189}$	$\frac{1}{190}$

but numbers composed only of *ones*, as **5, 7, 11, 13, 17, 19, 23,** &c., are usually called prime numbers ; the term composite is therefore restricted to numbers composed of 2s, 3s, 4s, and all higher numbers ; every rectangle is a composite number, and may very easily be resolved into its factors, after a careful performance of the exercises already prescribed, or by finding the numbers in the table of multiples, and the factors above and on the left

In class : What factors produce 14 ? 15 ? 16 ? 18 ? 20 ? 21 ? 22 ? 24 ? 25 ? 27 ? 28 ? 30 ? 32 ? 33 ? 35 ? 36 ? 40 ? 42 ? 44 ? 45 ? 48 ? 49 ? 50 ? 54 ? 55 ? 56 ? 60 ? 63 ? 64 ? 66 ? 70 ? 72 ? 77 ? 80 ? 81 ? 84 ? 88 ? 96 ? 99 ? 100 ? 108 ? 110 ? 120 ? 121 ? 132 ? 144 ?

Ans. 2 and 7 are the factors of 14 ; 3 and 5 are the factors of 15 ; 4 and 4, or 2 and 8, are the factors of 16, &c.

Of the formation of Rectangles, or Series whose terms are of two dimensions and variable increase.

Once 2=	what ?	Once 3=	what ?	Once 4=	what ?
twice 3=	?	twice 4=	?	twice 5=	?
3 × 4=	?	3 × 5=	?	3 × 6=	?
4 × 5=	?	4 × 6=	?	4 × 7=	?
5 × 6=	?	5 × 7=	?	5 × 8=	?
6 × 7=	?	6 × 8=	?	6 × 9=	?
7 × 8=	?	7 × 9=	?	7 × 10=	?
8 × 9=	?	8 × 10=	?	8 × 11=	?
9 × 10=	?	9 × 11=	?	9 × 12=	?
10 × 11=	?	10 × 12=	?	10 × 13=	?
11 × 12=	?	11 × 13=	?	11 × 14=	?

The student will please to mark these rectangles in pencil points upon the slate, giving to each its proper dimensions, length and breadth.

Let our readers now accompany us to the fort or navy yard, and see the order of the cannon balls piled in rectangular courses, forming series of terms similar to those above ; or it may be more agreeable to view the smiling hills of corn, or the fruitful orchard, laid out in similar rectangles.

191	192	193	194	195	196	197	198	199	200
$\frac{1}{191}$	$\frac{1}{192}$	$\frac{1}{193}$	$\frac{1}{194}$	$\frac{1}{195}$	$\frac{1}{196}$	$\frac{1}{197}$	$\frac{1}{198}$	$\frac{1}{199}$	$\frac{1}{200}$

Mercantile Numbers, upon the Prime Series.

One $ extends to 100 upon the line of cents; 2$ extends to what term? 3$ to what? 4$ to what? 5$ to what?

One lb. extends to 16 upon the line of ounces; 2 lbs. extend to what term? 3 lbs. to what? 4, 5, &c., to what?

One dwt. extends to 24 upon the line of grains; 2 dwts. extend to what term? 3 dwts. to what? 4, 5, 6, &c., to what?

One ton extends to 20 upon the line of cwts.; 2 tons extend to what term? 3 tons to what? 4, 5, 6, &c., to what?

One foot extends to 12 upon the line of inches; 2 feet extend to what term? 3 feet to what? 4, 5, 6, &c., to what?

One furlong extends to 40 upon the line of poles; 2 fur. extend to what term; 3, 4, 5, 6, &c., to what?

Of the Multiplication of High Numbers.

Rule

Case 1. When the multiplier does not exceed 12, place it under the multiplicand; multiply each rank successively, from the lowest to the highest; carry 1 for every 10 of each product, and set down the excess as in addition.

Proof. Reject 9s from the digits of each factor, and also from the product of the two excesses; the excess of 9s in this product will equal the same excess in the whole product.

Or, the product being a replication of the multiplicand, may be found by addition; also by subtracting the multiplicand repeatedly from the product, in the end nothing will be left; but it is a shorter proof to divide the product by the multiplier, and so to find the multiplicand.

Examples for the Slate.

1. Multiply 87654 by 2, and 98765 by 3.

		Excess.			Excess.
Md.	87654	= 3	Md.	98765	= 8
Mr.	2	= 2	Mr.	3	= 3
Pt.	175308	= 6	Pt.	296295	= 6

201	202	203	204	205	206	207	208	209	210
$\frac{1}{201}$	$\frac{1}{202}$	$\frac{1}{203}$	$\frac{1}{204}$	$\frac{1}{205}$	$\frac{1}{206}$	$\frac{1}{207}$	$\frac{1}{208}$	$\frac{1}{209}$	$\frac{1}{210}$

2. Multiply 123456789 by 2, and by 3.
3. Multiply 123456789 by 4, and by 5.
4. Multiply 123456789 by 6, and by 7.
5. Multiply 123456789 by 8, and by 9.
6. Multiply 123456789 by 10, and by 11.
7. Multiply 123456789 by 12, and by 13.

The class will now add together these 12 products.

8. Multiply 123456789 by 90, which is the sum of the twelve multipliers above; this product will equal the sum of the twelve products, if the work be true.

Rule.

Case 2. When the multiplier is a composite number; that is, the product of two or more factors; multiply by one of the factors, and that product by another, and the second product by a third, and so on.

In class: Resolve the following multipliers into their factors: viz., 16, 27, 35, 45, 55, 64, 72, 84, 88, 96, 108, 120, 132. And if help is required, the table of multiples is at hand; find the number in the table, and above, and on the left, are the factors.

Examples for the Slate.

1. Multiply 345678 by 16; that is, by 4×4, or 2×8.

Md. 345678	Md. 345678
Mr. 4×4=16.	Mr. 2×8=16.
1382712	691356
4	8
Pt. 5530848	Pt. 5530848

Note. These two operations prove each other.

2. Multiply 345678 by 27 and 35. } *Ans.* 55999836.
3. Multiply 345678 by 45 and 55. }
4. Multiply 345678 by 64 and 72. } *Ans.* 106468824.
5. Multiply 345678 by 84 and 88. }

211	212	213	214	215	216	217	218	219	220
$\frac{1}{211}$	$\frac{1}{212}$	$\frac{1}{213}$	$\frac{1}{214}$	$\frac{1}{215}$	$\frac{1}{216}$	$\frac{1}{217}$	$\frac{1}{218}$	$\frac{1}{219}$	$\frac{1}{220}$

6. Multiply 345678 by 96 and 108. }
7. Multiply 345678 by 120 and 132. } *Ans.* 157629168.

Note. In each of the answers above there are four products added together. The sum of the units of the twelve multipliers is 926; therefore, 345678 multiplied by 926 will give the sum of the twelve products. See Ex. 2, below.

Rule.

Case 3. When the multiplier is any number whatever: Make a product, first for the units, second for the tens, third for the hundreds, &c.; place the first figure of each product in the rank of its multiplier, and take the sum of all the products for the answer.

Examples for the Slate.

1. Multiply 78654 by 876, and by 543.

		Excess of 9s.		Excess of 9s.
Md.	78654=3		78654=3	
Mr.	876=3		543=3	
	471924			
	550578			
	629232			
Pt.	68900904=0		=?	

2. Multiply 345678 by 926. *Ans.* 320097828.
3. Multiply 345678 by 234. }
4. Multiply 45678 by 2345. } *Ans.* 188003562.
5. Multiply 45678 by 3456. }
6. Multiply 54678 by 4567. } *Ans.* 407577594.
7. Multiply 67891 by 12345. }
8. Multiply 45678 by 3570. } *Ans.* 1001184855.
9. Multiply 12345 by 10506. }
10. Multiply 54320 by 9875. } *Ans.* 666106570.
11. Multiply 9532 by 90178. }
12. Multiply 5849 by 8043. } *Ans.* 906620203.

Note. In order to disguise the answers, the sum of two answers are given. Some scholars are accused of *cobbling* answers.

221	222	223	224	225	226	227	228	229	230
$\frac{1}{221}$	$\frac{1}{222}$	$\frac{1}{223}$	$\frac{1}{224}$	$\frac{1}{225}$	$\frac{1}{226}$	$\frac{1}{227}$	$\frac{1}{228}$	$\frac{1}{229}$	$\frac{1}{230}$

Special Rules.

1. To multiply by 10, 100, 1000, &c., annex to the multiplicand the ciphers of the multiplier, always giving the right of the line to the units' rank.

2. Ciphers on the right of any multiplier may be joined to the multiplicand, multiplying the result by the significant figures.

3. To multiply by 9, 99, 999, &c., join to the multiplicand as many 0s as there are 9s; then subtract the multiplicand from the result.

4. When, in the multiplier, a composite number follows a factor of itself, as in 168, 16 follows 8, multiply the product of 8 by 2 of the rank above 8.

5. When the multiplier consists of two figures, one of which is a unit, as 18, 41, &c.; multiply by the other only, and add the multiplicand, according to the rank of the unit, whether first or second.

Examples for the Slate.

1. Multiply 365 by 10, and by 100.
2. Multiply 365 by 1000, and by 10000. } *Ans.* (See No. 11.)
3. Multiply 96 by 240, and by 300.
4. Multiply 96 by 360, and by 2400. } *Ans.* (See No. 12.)
5. Multiply 52 by 9, and by 99.
6. Multiply 52 by 999, and by 9999. } *Ans.* (See No. 13.)
7. Multiply 76 by 168, and by 189.
8. Multiply 76 by 147, and by 126. } *Ans.* (See No. 14.)
9. Multiply 286 by 18, and by 41.
10. Multiply 286 by 19, and by 61. } *Ans.* (See No. 15.)
11. Multiply 365 by 11110. *Ans.* 4055150.
12. Multiply 96 by 3300. *Ans.* 316800.
13. Multiply 52 by 11110, and subtract 4×52. *Ans.* 577512.
14. Multiply 76 by 630. *Ans.* 47880.
15. Multiply 286 by 139. *Ans.* 39754.

SIMPLE DIVISION.

To divide is to cut into equal sections or parts, to find how often any measure may be applied to a quantity.

The prime series, or line of numbers, to 2520 terms, extends along the head of these pages. The spaces occupied by the numbers may represent units of any species, or denomination; as men, dollars, acres, tons, miles, &c.

231	232	233	234	235	236	237	238	239	240
$\frac{1}{231}$	$\frac{1}{232}$	$\frac{1}{233}$	$\frac{1}{234}$	$\frac{1}{235}$	$\frac{1}{236}$	$\frac{1}{237}$	$\frac{1}{238}$	$\frac{1}{239}$	$\frac{1}{240}$

The measure of a quantity is that which divides it exactly. Every number used as a measure is *one ;* it divides the quantity into units of its own value.

The greatest measure that will divide a quantity is its half; but the half of 2520 is too large for beginners. Let us begin with the smaller measures.

Class Exercise upon the Line of Numbers.

1. How often will the measure 1 apply to 2520?

As the merchant applies his yard-stick to the cloth, saying *one*, *two*, *three*, *four*, &c., so the class, in succession, not in chorus, count the spaces upon the line of numbers; thus, one, two, three, four—Hold! the answer is known; it is 2520; because every number declares its own distance, in units, from the first point of the series.

2. How often will the measure 2 apply to 2520?

The class, calling 2 *one*, 4 *two*, 6 *three*, &c., proceed along the series of even numbers to the end.

Note. Repeating one term every second, the class will occupy 21 minutes upon this question. The time will be well employed.

3. How often will the measure 3 apply to 2520?

The class, calling 3 *one*, 6 *two*, 9 *three*, proceed to the end in 14 minutes.

After the same manner apply the following measures:

4. In 2520 how many 4s?
5. In 2520 how many 5s?
6. In 2520 how many 6s?
7. In 2520 how many 7s?
8. In 2520 how many 8s?
9. In 2520 how many 9s?
10. In 2520 how many 10s?
11. In 2520 how many 12s?

A common measure of two or more terms is that which will divide each of them exactly. Therefore,

2 is a common measure of every even number.

3 is a common measure of every 3d term in the series.

4, of every 4th term; 5, of every 5th term;

6, of every 6th term; 7, of every 7th term;

8, of every 8th term; 9, of every 9th term;

10, of every 10th term; and so of all others.

The dividend, divisor, and quotient, are terms or words applied to quantities in division. The former two are

241	242	243	244	245	246	247	248	249	250
$\frac{1}{241}$	$\frac{1}{242}$	$\frac{1}{243}$	$\frac{1}{244}$	$\frac{1}{245}$	$\frac{1}{246}$	$\frac{1}{247}$	$\frac{1}{248}$	$\frac{1}{249}$	$\frac{1}{250}$

given, the latter is required. The dividend and quotient are of the same species; the divisor is also the same in parcels, of which the quotient is one.

Dividuals, subtrahends, and remainders, are terms which result from the operation.

Because dividend÷divisor=quotient÷1, the terms are proportionals; thus, dd : dr : : qt : 1; or dd : qt : : dr : 1.

When the divisor does not exactly measure the dividend, but leaves a remainder at the last, the quotient is completed by writing the divisor under the remainder: thus, $18 \div 4 = 4\frac{2}{4}$; $21 \div 5 = 4\frac{1}{5}$, &c.

Mercantile Numbers upon the Prime Series.

100 cents count 1 upon the line of dollars; 200 cents count how many? 300 cts., how many? 400, how many?

16 ounces count 1 upon the line of pounds; 32 oz. count how many? 48, how many? 64, 80, &c., how many?

24 grains count 1 upon the line of dwts.; 48 grs. count how many? 72 grs., how many? 96, 120, &c., how many?

20 cwt. count 1 upon the line of tons; 40 cwt. count how many? 60 cwt., how many? 80, 100, &c., how many?

12 inches count 1 upon the line of feet; 24 in. count how many? 36 in., how many? 48, 60, &c., how many?

40 poles count 1 upon the line of furlongs; 80 poles count how many? 120, how many? 160, 200, &c., how many?

144 sq. inches count 1 upon the line of sq. feet; 288 sq. in. count how many? 432, how many? 576, &c., how many?

8 quarts count 1 upon the line of pecks; 16 qts. count how many? 24, 32, 40, &c., how many?

60 minutes count 1 upon the line of hours; 120 min. count how many? 180, how many? 240, 300, &c., how many?

30 degrees count 1 upon the line of signs; 60 deg. count how many? 90, 120, 150, &c., how many?

251	252	253	254	255	256	527	258	259	260
$\frac{1}{251}$	$\frac{1}{252}$	$\frac{1}{253}$	$\frac{1}{254}$	$\frac{1}{255}$	$\frac{1}{256}$	$\frac{1}{257}$	$\frac{1}{258}$	$\frac{1}{259}$	$\frac{1}{260}$

Statement of the Terms in Division.

It is usual to place the divisor, great or small, on the left of the dividend; if great, the quotient takes the right of the dividend, and the work appears in a succession of subtractions below it; if small, the quotient is set below, and the work is more mental. The former is called Long, and the latter

Short Division.

Rule. When the divisor does not exceed 12, the first, or first two or three figures will contain it a certain number of times, and mostly leave a remainder; set down the number of times as quotient; and join to the remainder (if any, if not, take) the next figure for a new dividual, which will also contain the divisor and leave a remainder, with which proceed as with the former; and so to the end. Complete the quotient by annexing to it the final remainder with the divisor subscribed.

Proof. The product of the complete quotient and divisor will equal the dividend, if the work be true.

Examples for the Slate.

1. Divide 987654321 by 2 and by 3.

2)987654321		3)987654321
$493827160\frac{1}{2}$	Quotient.	329218107
2		3
987654321	Proof.	987654321

2. Divide 987654321 by 3, and by 4.
3. Divide 987654321 by 5, and by 6.
4. Divide 987654321 by 7, and by 8.
5. Divide 987654321 by 9, and by 10.
6. Divide 987654321 by 11, and by 12.
7. Divide 123456789 by 4, 5, and 6.
8. Divide 123456789 by 7, 8, 9, 10, 11, and 12.

Note. When the divisor is a composite number, use its

261	262	263	264	265	266	267	268	269	270
$\frac{1}{261}$	$\frac{1}{262}$	$\frac{1}{263}$	$\frac{1}{264}$	$\frac{1}{265}$	$\frac{1}{266}$	$\frac{1}{267}$	$\frac{1}{268}$	$\frac{1}{269}$	$\frac{1}{270}$

factors, if small, successively, under the rule for Short Division. In such operations, the first remainder, if the only one, will be the true one; but generally, the last remainder $\times$last divisor but one$+$its remainder. This result$\times$last divisor but two$+$its remainder, &c., will result in the true remainder.

But the rule for completing every quotient, as above, is to be preferred, as obviating all difficulty in the case of remainders.

Examples for the Slate.

1. Divide 933330645 by $27=3\times9$.

$$27=9\times3)933330645$$
$$9)311110215$$

Proof by multiplication. $34567801\frac{6}{9}$

2. Divide 12098730 by 35. *Ans.* 345678.
3. Divide 15555510 by 45. *Ans.* 345678.
4. Divide 19012290 by 55. *Ans.* 345678.
5. Divide 22123392 by 56. *Ans.* $395060\frac{32}{56}$.
6. Divide 24888816 by 72. *Ans.* 345678.
7. Divide 30419664 by 88. *Ans.* 345678.
8. Divide 85492765 by 96. *Ans.* $890549\frac{61}{96}$.
9. Divide 75798432 by 110. *Ans.* $689076\frac{72}{110}$.
10. Divide 59846594 by 132. *Ans.* $453383\frac{38}{132}$.

Long Division.

Long Division exhibits the process of the operation more fully, and is therefore more simple than Short Division. It is used in the case of large divisors, but is equally applicable to the smallest.

Rule. Seek how often one or two places on the left of the dividend will contain the highest rank of the divisor, allowing for the increase to be carried; multiply the divisor by the figure found, and subtract the product from the first dividual; join the next figure of the dividend to the remainder for the second dividual, which divide as the

271	272	273	274	275	276	277	278	279	280
$\frac{1}{271}$	$\frac{1}{272}$	$\frac{1}{273}$	$\frac{1}{274}$	$\frac{1}{275}$	$\frac{1}{276}$	$\frac{1}{277}$	$\frac{1}{278}$	$\frac{1}{279}$	$\frac{1}{280}$

former; continue the same process to the end, annexing the last remainder to the quotient, with the divisor written under it.

Proof. The method proposed in the last rule is general, or thus: subtract the last remainder from the dividend, and divide the residue by the quotient; or thus: the sum of all the subtrahends, with the remainder, taken in order, will equal the dividend.

Examples for the Slate.

1. Divide 846372 by 480.

```
480)846372(1763 132/480   =   1763 11/40
    480...                    40×12=480
    ----                      -----
    3663                      70531
    3360                         12
    ----                      ------
     3037                     846372 Proof.
     2880                     ======
     ----
      1572
      1440
      ----
       132
       ===
```

2. Divide 12345678 by 6543. *Ans.* 1886, rem. 5580.
3. Divide 7416496 by 2468. *Ans.* 3005, rem. 156.
4. Divide 4813729 by 6496. *Ans.* 741, rem. 193.
5. Divide 3508901 by 4813. *Ans.* 729, rem. 224.
6. Divide 45757390 by 9365. *Ans.* 4886.
7. Divide 33809250 by 4575. *Ans.* 7390.
8. Divide 31265000 by 9250. *Ans.* 3380.
9. Divide 3160708 by 3508. *Ans.* 901.
10. Divide 246480 by 708. *Ans.* $348\frac{96}{708}$.

Note 1. Ciphers on the right of a divisor may be rejected, omitting, also, as many places on the right of the dividend, which must be joined to the last remainder.

1. Divide 3704196 by 20. *Ans.* $185209\frac{16}{20}$.
2. Divide 31086901 by 7100. *Ans.* $4378\frac{3101}{7100}$.

281	282	283	284	285	286	287	288	289	290
$\frac{1}{281}$	$\frac{1}{282}$	$\frac{1}{283}$	$\frac{1}{284}$	$\frac{1}{285}$	$\frac{1}{286}$	$\frac{1}{287}$	$\frac{1}{288}$	$\frac{1}{289}$	$\frac{1}{290}$

3. Divide 7380964 by 23000. *Ans.* $320\frac{20964}{23000}$.
4. Divide 2304109 by 5800. *Ans.* $397\frac{1509}{5800}$.
5. Divide 83016572 by 240. *Ans.* $345902\frac{92}{240}$.

Note 2. When the divisor is 10, 100, 1000, &c., reject the ciphers, and cut off for a remainder as many places on the right of the dividend, under which subscribe the divisor; the unit will divide no number.

Examples for the Slate.

1. Divide 345678 by 10. *Ans.* $34567\frac{8}{10}$.
2. Divide 345678 by 100. *Ans.* $3456\frac{78}{100}$.
3. Divide 345678 by 1000. *Ans.* $345\frac{678}{1000}$.
4. Divide 345678 by 10000. *Ans.* $34\frac{5678}{10000}$.
5. Divide 345678 by 100000. *Ans.* $3\frac{45678}{100000}$.
6. Divide 345678 by 1000000. *Ans.* $\frac{345678}{1000000}$.

Note. These last extend the decimal notation to the right of unity, and exhibit a series called tenths, hundredths, thousandths, ten thousandths, &c., of a unit. And it is not necessary to subscribe the denominators or divisors, for the numerator shows the number of ciphers joined to the unit, in forming the denominator; thus, 345678 is the number undivided, and the successive divisions remove the units' place onward to the left, until at length it vanishes out of the number. Therefore the above answers may be written as follows, viz.: 34567.8, 3456.78, 345.678, 34.5678, 3.45678, and .345678, in all of which the period is the units' point.

FEDERAL MONEY AND DECIMALS.

The *reading* of fractions is now in order. Therefore, beginning at $\frac{1}{1}$, page 7, second line, the class will repeat the terms of the series of ordinal numbers, in succession; viz., first, second, third, fourth, fifth, sixth, &c., to the 2520th term.

These ordinals are denominators. *First* is the first *part*, as *one* is the first multiple of *one*; *half* is used for second part, and *quarter* for fourth part. It is unnecessary to repeat *one* so often, as numerator.

291	292	293	294	295	296	297	298	299	300
$\frac{1}{291}$	$\frac{1}{292}$	$\frac{1}{293}$	$\frac{1}{294}$	$\frac{1}{295}$	$\frac{1}{296}$	$\frac{1}{297}$	$\frac{1}{298}$	$\frac{1}{299}$	$\frac{1}{300}$

In reading or repeating the terms of the fractional series, the class finds the decimal numbers and reciprocals in the following order; namely,

Multiples, 1, 10, 100, 1000, 10000, &c.
Parts, $\frac{1}{1}$, $\frac{1}{10}$, $\frac{1}{100}$, $\frac{1}{1000}$, $\frac{1}{10000}$, &c.

And because, in numerating, the higher periods and ranks are first read, the *multiples* are reversed and brought into line with the *parts*, thus:

10000, 1000, 100, 10, 1, $\frac{1}{10}$, $\frac{1}{100}$, $\frac{1}{1000}$, $\frac{1}{10000}$, &c.

But, as one contrivance may be as good as another, the same fractions may be expressed in a more convenient form; namely,

thus, 10000, 1000, 100, 10, 1, .1, .01, .001, .0001, &c.

And since the 0s are but counters of the distance of each significant figure from the units' place, we can remove them; for it is plain that the 1s will perform this office for each other:

thus, 1 1 1 1 1 .1 1 1 1

Now if these 1s were moveable, we could easily press them nearer each other; thus, 11111.1111; which makes no change in their relative position.

Other digits, also, instead of the 1s, might occupy the several ranks; because nine units is the full complement of each rank; therefore any selection of *one* digit from each space in the series below, or a 0 to fill blank places, is allowed by the law of the notation.

1000000s.	100000s.	10000s.	1000s.	100s.	10s.	. units.	10ths.	100ths.	1000ths.	10000ths.	100000ths.
123	234	345	456	567	678	789	891	912	123	234	345
456	567	678	789	891	912	123	234	345	456	567	678
789	891	912	123	234	345	456	567	678	789	891	912

This diagram presents two series of numbers; the one having for its denominator an undivided unit, the other a unit divided into one hundred thousand parts.

301	302	303	304	305	306	307	308	309	310
$\frac{1}{301}$	$\frac{1}{302}$	$\frac{1}{303}$	$\frac{1}{304}$	$\frac{1}{305}$	$\frac{1}{306}$	$\frac{1}{307}$	$\frac{1}{308}$	$\frac{1}{309}$	$\frac{1}{310}$

The divisions of Federal money are the same as the decimal notation, the dollar being the unit; viz.,

10 mills, or 1000ths make 1 cent, or 100th.
10 cents, or 100ths, make 1 dime, or 10th.
10 dimes, or 10ths, make 1 dollar or unit.
10 dollars, or 1s, make one eagle, or ten.

The eagle and dime are not named in accounts, and the mill is too small to be regarded, unless the number of them is equal to $\frac{1}{4}$ or $\frac{1}{2}$ cent.

The dimes and cents occupy two ranks, and when either place is without a significant figure, a cipher ought to fill it.

The eagle, $\frac{1}{2}$ eagle, and $\frac{1}{4}$ eagle are gold coins.

The dollar, $\frac{1}{2}$ dollar, $\frac{1}{4}$ dollar, dime and $\frac{1}{2}$ dime are silver. The relative value of the gold coins to the silver is fixed by law at 1, in weight, to 16.

The units' point should not be omitted before cents under the title of dollars; under its own title every cent or other denomination is one; under the title of dollars every cent is $\frac{1}{100}$, or .01.

Of the Reading of Decimal Fractions.

Point, five (.5), reads five tenths.

Point, two, five (.25), reads 25 hundredths.

Point, three, seven, five (.375), reads 375 thousandths.

Point, 6, 8, 7, 5 (.6875), reads 6875 ten thousandths.

Point, 5, 6, 8, 7, 5 (.56875), reads 56 thousand 875 hundred thousandths.

Of the Reduction of Federal Money.

Let the units' point always stand on the right of unity.

To change a low title to a higher one requires the removal of the units' point toward the higher.

Examples for the Slate.

45678. mills make 4567.8 cents.
4567.8 cents make 456.78 dimes.
456.78 dimes make 45.678 dollars.
45.678 dollars make 4.5678 eagles.

311	312	313	314	315	316	317	318	319	320
$\frac{1}{311}$	$\frac{1}{312}$	$\frac{1}{313}$	$\frac{1}{314}$	$\frac{1}{315}$	$\frac{1}{316}$	$\frac{1}{317}$	$\frac{1}{318}$	$\frac{1}{319}$	$\frac{1}{320}$

To change a higher title to a lower one, requires the removal of the units' point toward the lower.

Examples for the Slate.

4.5678 eagles make 45.678 dollars.
45.678 dollars make 456.78 dimes.
456.78 dimes make 4567.8 cents.
4567.8 cents make 45678. mills.

Since 0s are used only as counters of the distance of significant figures from the units' rank, they make no change in the value of figures which are between them and that rank :

Thus, 0005. 005. 05. 5. are of the simple value of five.
and .5, .50, .500, .5000, are of the value of 5 tenths.
For the same reason 0s give notice of a change in the value of such figures as are outside of them.

Thus, 5. 50. 500. 5000. increase in the ratio of 1 to 10.
and .5, .05, .005, .0005, decrease in the ratio of 10 to 1.

But, though .5, .50, .500, &c., are of the same value, there is an increase in the number of parts, and a corresponding decrease in the value of each part.

The terms of a fraction are the numerator and denominator. Both terms may be multiplied by the same number, without changing the value of the fraction; for if the number of the parts be so increased, the value of each part is diminished in the same ratio. For the same reason, both terms may be divided by the same number.

Examples for the Slate.

Multiply the terms—		Divide the terms—	
of $\frac{5}{10}$ by 10.	*Ans.* $\frac{50}{100}$.	of $\frac{500}{1000}$ by 10.	*Ans.* $\frac{50}{100}$.
of $\frac{50}{100}$ by 10.	*Ans.* $\frac{500}{1000}$.	of $\frac{50}{100}$ by 10.	*Ans.* $\frac{5}{10}$.
of $\frac{1}{2}$ by 2.	*Ans.* $\frac{2}{4}$.	of $\frac{8}{16}$ by 2.	*Ans.* $\frac{4}{8}$.
of $\frac{2}{4}$ by 2.	*Ans.* $\frac{4}{8}$.	of $\frac{4}{8}$ by 2.	*Ans.* $\frac{2}{4}$.
of $\frac{4}{8}$ by 2.	*Ans.* $\frac{8}{16}$.	of $\frac{2}{4}$ by 2.	*Ans.* $\frac{1}{2}$.

321	322	323	324	325	326	327	328	329	330
$\frac{1}{321}$	$\frac{1}{322}$	$\frac{1}{323}$	$\frac{1}{324}$	$\frac{1}{325}$	$\frac{1}{326}$	$\frac{1}{327}$	$\frac{1}{328}$	$\frac{1}{329}$	$\frac{1}{330}$

Addition of Federal Money.

The parts of a dollar being decimal, any operation performed upon them will apply to miles, tons, yards, gallons, or any other denomination similarly divided.

RULE. Place units under units, inserting the units' point; then the 10ths, 100ths, and other ranks will fall into their proper order. Begin at the right or lowest rank; add and carry one for every 10, as in simple addition, setting the units' point in the sum under those of the given numbers.

Examples for the Slate.

.0625	.5625	123.45	24.56
.125	.625	67.891	18.57
.1875	.6875	2.34	34.68
.25	.75	567.8	5.09
.3125	.8125	90.1234	6.93
.375	.875	5.678	8.62
.4375	.9375	9.13	73.37
.5	1.	91.3	2.8
$2.2500	$.	$	

1. Merchants who keep exact accounts, enter halves and quarters of a cent in their money columns; and in adding, the clerk, calling each ½ cent ²⁄₄, sums them up, and reduces the sum to cents, dividing by 4.

$	cts.	$	cts.	$	cts.	$	cts.
347	37½	123	12½	987	93¾	321	56¼
916	12½	234	18¾	876	87½	210	50
784	56¼	345	25	765	81¼	101	43¾
812	43¾	456	31¼	654	75	12	37½
218	25	567	37½	543	68¾	123	31¼
954	50	678	43¾	432	62½	234	25
$	?	$	?	$	?	$	?

2. What sum, under the title of dollars, will 4567 cents 5678 cents, 6789 cents, 1234 cents, 2345 cents, and 34567 cents amount to? *Ans.* $551.80.

331	332	333	334	335	336	337	338	339	340
1/331	1/332	1/333	1/334	1/335	1/336	1/337	1/338	1/339	1/340

3. What sum in dollars will 3456 mills, 4567 mills, 5678 mills, 6789 mills, 1234 mills, 2345 mills, and 34567 mills amount to? *Ans.* $58.636.

4. What is the amount of $2.50, $3.75, $5.12½, $6.25, $7.43¾, $8.87½, and $56.11¼? *Ans.* $90.05.

5. What is the value, under the title of dollars, of 567$, 567 dimes, 567 cents, 567 mills? *Ans.* $629.937.

6. What is the worth, in dollars, of 678 eagles, 678 dollars, 678 dimes, 678 cents, and 678 mills?
Ans. $7533.258.

7. How much do the following sums amount to in dollars, viz., 34560 mills, $6.93, $35.75, $69.18¾, $356.18¾, $85.43¾, and 54289 cents? *Ans.* $1131.84¼.

8. What did 2 lbs of butter at 31¼ cents each, 2 lbs. of cheese at 18¾ cents each, 6 lbs. of coffee at 11¼ cents each and 40 lbs. of sugar, costing in all $2.62½, amount to?
Ans. $4.30.

Subtraction of Federal Money and Decimals.

Rule. Set the less number under the greater, with units under units, tenths under tenths, not omitting the units' point; proceed as in Simple Subtraction, setting the units' point in the remainder, in a line with the points of the given numbers.

Note. If there be parts of a cent in the given terms, reduce them to the same name; take the lower numerator from the upper, if possible; if not, increase the upper by a unit reduced to the same name; subtract, and carry 1 to the next lower place.

Examples for the Slate.

From	$232.56	From	$34.75	From	$2.12½
Take	129.857	Take	29.18¾	Take	1.68¾
	$102.703		$5.56¼		$?
From	$5.81¼	From	$4.86	From	$12.25
Take	3.93¾	Take	1.57	Take	6.78
	$?		$?		$?

341	342	343	344	345	346	347	348	349	350
$\frac{1}{341}$	$\frac{1}{342}$	$\frac{1}{343}$	$\frac{1}{344}$	$\frac{1}{345}$	$\frac{1}{346}$	$\frac{1}{347}$	$\frac{1}{348}$	$\frac{1}{349}$	$\frac{1}{250}$

7. From 45678 cents take 45678 mills. *Ans.* $411.102.

8. From 456 dimes take 456 cents. *Ans.* $41.04.

9. From 4567 dimes take 4567 mills. *Ans.* $452.133.

10. From $5 take 93¾ cts., 81¼ cts., 56¼ cts., 43¾ cts., 18¾ cts., and 6¼ cts. *Ans.* $2.

11. From 1000 dollars subtract $256.25, $258.75, $285.87½, and 124.62½, in succession. *Ans.* $74.50.

12. From $750.62½ take $75.06¼. *Ans.* $675.56¼.

13. From 567 dollars+567 dimes+567 cents+567 mills, take $375.62½; what will remain? *Ans.* $254.312.

14. From $2345 take 678 dollars+678 dimes+678 cents, +678 mills; what will be the residue? *Ans.* $1591.742.

15. From the sum of $3.50, $4.37½, $5.25, and $6.12½, take the sum of $2.43¾, $3.31¼, 4.18¾, and $5.06¼; what will then be left? *Ans.* $4.25.

16. From $2345 take 2345 mills. *Ans.* 2342.655.

Multiplication of Federal Money and Decimals.

When, in multiplication, one of the factors is less than a unit, the product will be in the same ratio, less than the other factor:

for, 1 : mr : : md : pt; or, 1 : md : : mr : pt.

But every fraction is less than a unit; for it is a unit, or some number, divided by one of higher rank; and to divide one of the factors is equivalent to dividing the product.

Rules. The product of two ranks on the same side of unity is more remote from unity than either of the factors. The product of two ranks, equidistant from and on opposite sides of unity, is nearer to unity than either of the factors. The product of any two opposite ranks is a mean rank.

Units×units=units, units×tenths=tenths.
units×100ths=100ths, tens×tenths=units.
tens×100ths=10ths, 10ths×10ths=100ths.
tenths×100ths=1000ths.

351	352	353	354	355	356	357	358	359	360
$\frac{1}{351}$	$\frac{1}{352}$	$\frac{1}{353}$	$\frac{1}{354}$	$\frac{1}{355}$	$\frac{1}{356}$	$\frac{1}{357}$	$\frac{1}{358}$	$\frac{1}{359}$	$\frac{1}{360}$

Table of Decimal Multiplication.

Let the oblique cross be read *times*, when the multiplier is a multiple of 1; but when the multiplier is a fraction the cross reads *of*, or *parts of*, and the parallels (=) *equal* or *equals.*

	u 10ths.		u 10ths.		u 10ths.
2× 2 tenths=	.4	3× 9 tentns=	2.7	6 × 6 tenths=	3.6
2× 3 "	= .6	3×10 "	=3.	6 × 7 "	= 4.2
2× 4 "	= .8	4× 4 "	=1.6	6 × 8 "	= 4.8
2× 5 "	=1.	4× 5 "	=2.	6 × 9 "	= 5.4
2× 6 "	=1.2	4× 6 "	=2.4	6 ×10 "	= 6.
2× 7 "	=1.4	4× 7 "	=2.8	7 × 7 "	= 4.9
2× 8 "	=1.6	4× 8 "	=3.2	7 × 8 "	= 5.6
2× 9 "	=1.8	4× 9 "	=3.6	7 × 9 "	= 6.3
2×10 "	=2.	4×10 "	=4.	7 ×10 "	= 7.
3× 3 "	= .9	5× 5 "	=2.5	8 × 8 "	= 6.4
3× 4 "	=1.2	5× 6 "	=3.	8 × 9 "	= 7.2
3× 5 "	=1.5	5× 7 "	=3.5	8 ×10 "	= 8.
3× 6 "	=1.8	5× 8 "	=4.	9 × 9 "	= 8.1
3× 7 "	=2.1	5× 9 "	=4.5	9 ×10 "	= 9.
3× 8 "	=2.4	5×10 "	=5.	10×10 "	=10.

2 tenths×.2=.04	3 tenths×.7=.21	5 tenths×.8=.4
2 " ×.3=.06	3 " ×.8=.24	5 " ×.9=.45
2 " ×.4=.08	3 " ×.9=.27	6 " ×.6=.36
2 " ×.5=.1	4 " ×.4=.16	6 " ×.7=.42
2 " ×.6=.12	4 " ×.5=.2	6 " ×.8=.48
2 " ×.7=.14	4 " ×.6=.24	6 " ×.9=.54
2 " ×.8=.16	4 " ×.7=.28	7 " ×.7=.49
2 " ×.9=.18	4 " ×.8=.32	7 " ×.8=.56
3 " ×.3=.09	4 " ×.9=.36	7 " ×.9=.63
3 " ×.4=.12	5 " ×.5=.25	8 " ×.8=.64
3 " ×.5=.15	5 " ×.6=.3	8 " ×.9=.72
3 " ×.6=.18	5 " ×.7=.35	9 " ×.9=.81

From the foregoing remarks and exercises is deduced the following

RULE. Proceed in this as in Simple Multiplication, and point off in the product as many ranks on the right of the units' rank as there are so pointed in both the factors.

Contraction. Invert the order of the multiplier, setting the units' figure under the lowest rank to be retained in the product; reject of the upper factor all the ranks to the right of each multiplying figure, adding what should be carried from the places omitted; viz., 1 for 5, 2 for 15,

361	362	363	364	365	366	367	368	369	370
$\frac{1}{361}$	$\frac{1}{362}$	$\frac{1}{363}$	$\frac{1}{364}$	$\frac{1}{365}$	$\frac{1}{366}$	$\frac{1}{367}$	$\frac{1}{368}$	$\frac{1}{369}$	$\frac{1}{370}$

3 for 25, making a judicious average of the loss and gain in the rejected figures; and set the first figures of the products in a vertical line for addition.

This contraction is necessary in preparing tabular numbers.

Examples for the Slate.

1. Multiply 27.14986 by 92.41035.

Common form.	Contracted form.
27.14986	27.14986
92.41035	53014.29
13574930	24434874
8144958	542997
27149860	108599
10859944	2715
5429972	81
24434874	14
2508.9280650510	2508.9280

The operation is here contracted to 4 places.

2. Multiply 79.347 by 23.15. *Ans.* 1836.88305.
3. Multiply .63478 by .8264. *Ans.* .524582192.
4. Multiply .385746 by .0463. *Ans.* .00178600398.
5. Multiply 1.5708 by .5. *Ans.* .7854.
6. Multiply .125$ by 64. *Ans.* 8.
7. Multiply .0625$ by 128 *Ans.* 8.
8. Multiply .1875 by 40. *Ans.* 7.5.
9. Multiply .25 by 256. *Ans.* 64.
10. Multiply .3125 by 360. *Ans.* 112.5
11. Multiply .375 by 72. *Ans.* 27.
12. Multiply .4375 by 256. *Ans.* 112.
13. Multiply .0625, .125, .1875, .25, .3125, .375, .4375, and .5, each by 20, and tell the sum of all the products.
14. Multiply .5625, .625, .6875, .75, .8125, .875, .9375, and $1, each by 20, and tell the sum of all the products.
15. Multiply 2.25+6.25 by 20; this product shall equal

371	372	373	374	375	376	377	378	379	380
$\frac{1}{371}$	$\frac{1}{372}$	$\frac{1}{373}$	$\frac{1}{374}$	$\frac{1}{375}$	$\frac{1}{376}$	$\frac{1}{377}$	$\frac{1}{378}$	$\frac{1}{379}$	$\frac{1}{380}$

the sum of the products in the last two examples, if the work be true.

16. Multiply 75 dols. 75 dimes, 75 cts., and 75 mills, each by 64, and tell the amount of all the products.

17. Multiply 83.325 dollars by 64; this product shall equal the amount of all the products in the last example, if the work be true.

18. Multiply 1.06 by 1.06, and the product by 1.06; also this product by 1.06; and invert the multiplier in the third operation, retaining only 6 decimal places.
Last product, 1.262477.

19. Multiply 1.05 by 1.05, and the product by 1.05; also this product by 1.05; and invert the multiplier in the third operation, contracting the decimal to six places.
Last product, 1.215506.

20. Multiply 1.04 by 1.04, and the product by 1.04; also this product by 1.04; and invert the multiplier in the third operation, contracting the decimal to 6 places.
Last product, 1.169858.

21. Multiply 1.03 by 1.03, and the product by 1.03; also this product by 1.03; and invert the multiplier in the third operation, contracting the decimal to 6 places.
Last product, 1.125509.

22. Multiply 1.055 by 1.055, and the product by 1.055; also this product by 1.055; and invert the order of the multiplier in the second and third operations, contracting the decimal to 6 places. Last product, 1.238824.

Division of Federal Money and Decimals.

Since dd : dr : : qt : 1; if dividend be greater than divisor, quotient will be greater than 1; if equal, equal; and if less, less. But all these cases occur in this division.

One of the objects which this rule has in view is to reduce common fractions to the decimal form. And when the numerator or dividend is less than the denominator or divisor, it provides means of increasing the number of the parts by lessening the value of each.

Thus, in the fraction $\frac{1}{2}$, 1 is to be divided by 2, which

381	382	383	384	385	386	387	388	389	390
$\frac{1}{381}$	$\frac{1}{382}$	$\frac{1}{383}$	$\frac{1}{384}$	$\frac{1}{385}$	$\frac{1}{386}$	$\frac{1}{387}$	$\frac{1}{388}$	$\frac{1}{389}$	$\frac{1}{390}$

cannot be done until 1 is reduced to parts sufficiently numerous.

Let 1 be reduced to 10ths; then there are 10 parts to be divided by 2; each of the parts is $\frac{1}{10}$, and 5 is the half of them; so that $\frac{5}{10}$, or .5, is the decimal form of $\frac{1}{2}$.

But there is a formula, or Rule, for these reductions.

Rule. Bring the numerator to the right of the denominator, using the parenthesis; annex 0s successively, as far as may be required, and divide, setting the units' point before the figures resulting from the annexed ciphers.

Some quotients repeat the same figure, others repeat several figures, and never terminate.

Examples for the Slate.

$\frac{1}{2}$)1.0	$\frac{1}{3}$)1.000	$\frac{1}{4}$)1.00	$\frac{1}{6}$)1.00000
.5	.333$\frac{1}{3}$	.25	.16666$\frac{4}{6}$

Reduce $\frac{1}{5}$ to a decimal. Reduce $\frac{3}{4}$ to a decimal.
Reduce $\frac{1}{7}$ to a decimal. Reduce $\frac{3}{5}$ to a decimal.
Reduce $\frac{1}{8}$ to a decimal. Reduce $\frac{3}{6}$ to a decimal.
Reduce $\frac{1}{9}$ to a decimal. Reduce $\frac{3}{7}$ to a decimal.
Reduce $\frac{1}{11}$ to a decimal. Reduce $\frac{3}{8}$ to a decimal.
Reduce $\frac{1}{12}$ to a decimal. Reduce $\frac{4}{9}$ to a decimal.
Reduce $\frac{1}{8}$, $\frac{3}{8}$, $\frac{5}{8}$, and $\frac{7}{8}$ to equivalent decimals.

Reduce $\frac{1}{16}$, $\frac{3}{16}$, $\frac{5}{16}$, $\frac{7}{16}$, $\frac{9}{16}$, $\frac{11}{16}$, $\frac{13}{16}$, and $\frac{15}{16}$ to equivalent decimals. And since 16 is a large divisor, place the quotient on the right, and proceed as in Long Division.

Another important use to be made of this Rule is to find the rate of 1 from the given rate per cent. This requires nothing more than the removal of the units' point two places to the left; and since in the usual rates per cent. there is only one figure, a cipher must be used to fill the blank.

Examples for the Slate.

What will be the rate per unit at the following rates per cent., or per 100:

1. At 10 per cent., or per 100? *Ans.* $\frac{10}{100}$=.10.

391	392	393	394	395	396	397	398	399	400
$\frac{1}{391}$	$\frac{1}{392}$	$\frac{1}{393}$	$\frac{1}{394}$	$\frac{1}{395}$	$\frac{1}{396}$	$\frac{1}{397}$	$\frac{1}{398}$	$\frac{1}{399}$	$\frac{1}{400}$

2. At 9 per cent., or per 100? *Ans.* $\frac{9}{100}$=.09.
3. At 8 per cent., or per 100? *Ans.* $\frac{8}{100}$=.08.
4. At 7 per cent., or per 100? *Ans.* $\frac{7}{100}$=.07.
5. At 6 per cent., or per 100? *Ans.* $\frac{6}{100}$=.06.
6. At 5 per cent., or per 100? *Ans.* $\frac{5}{100}$=.05.
7. At 4 per cent., or per 100? *Ans.* $\frac{4}{100}$=.04.
8. At 3 per cent., or per 100? *Ans.* $\frac{3}{100}$=.03.
9. At 2 per cent., or per 100? *Ans.* $\frac{2}{100}$=.02.
10. At 1 per cent., or per 100? *Ans.* $\frac{1}{100}$=.01.

And since $\frac{1}{2}$= .5, $6\frac{1}{2}$=6.5 per c., and .065 per unit;
also, $\frac{1}{4}$=.25, $6\frac{1}{4}$=6.25 per c., and .0625 per unit;
and $\frac{3}{4}$=.75, $5\frac{3}{4}$=5.75 per c., and .0575 per unit.

Thus the same exercise may be extended to a class of rates made up of multiples and parts of a unit.

Set down the commission, interest, or insurance upon 1 dollar, at the following rates per cent.

$6\frac{1}{2}$, $6\frac{1}{4}$, $5\frac{3}{4}$, $5\frac{1}{2}$, $5\frac{1}{4}$, $4\frac{3}{4}$, $4\frac{1}{2}$, $4\frac{1}{4}$, $3\frac{3}{4}$, $3\frac{1}{2}$, $3\frac{1}{4}$, $2\frac{3}{4}$, $2\frac{1}{2}$, $2\frac{1}{4}$, $1\frac{3}{4}$, $1\frac{1}{2}$, $1\frac{1}{4}$, $\frac{3}{4}$, $\frac{1}{2}$, $\frac{1}{4}$.

From what has preceded, the following general rule is easily deduced.

RULE. Give to the dividend always as many decimal places as there are in the divisor, by annexing ciphers, and as many more as the case may require; proceed as in simple division, and point off in the quotient as many fractional ranks as the dividend has more than the divisor, supplying the defect, if any, with ciphers after the units' point.

To contract a long division. Take as many places of the divisor as you desire to be in the quotient,* and divide as usual; then instead of joining the figures of the dividend to the remainders, leave out the figures of the divisor, adding for the increase of the figures omitted, as taught in contracted multiplication.

---

* If the divisor have not so many figures, find a part of the quotient before the contraction commence.

401	402	403	404	405	406	407	408	409	410
$\frac{1}{401}$	$\frac{1}{402}$	$\frac{1}{403}$	$\frac{1}{404}$	$\frac{1}{405}$	$\frac{1}{406}$	$\frac{1}{407}$	$\frac{1}{408}$	$\frac{1}{409}$	$\frac{1}{410}$

Examples for the Slate.

1. Divide 2508.9280650510 by 27.14986 in the common form, and also contracted to 5 places.

```
27.14986)2508.9280650510(92.41035
         24434874······
         --------
          6544066
          5429972
          --------
          11140945
          10859944
          --------
            2810010
            2714986
            -------
             9502451
             8144958
             --------
             13574930
             13574930
             --------
```

Contracted to 5 quotient figures.

```
27.14986)2508.9280650510(92.410
         244349
         ------
          6543
          5430
          ----
          1113
          1085
          ----
            28
            27
            --
             1
```

NOTE.—The four questions, Nos. 5, 6, 7, 8, on p. 48, have in view the formation of a decreasing series, showing the several worths of $1 for the future years, from 1 to 20. The same series may, however, be more readily found by involution; for if $1 be divided by its amount for a year, at any given rate, the powers of that quotient will be the same as the quotients demanded in the four questions referred to. The learner may be excused from seeking any more than the 1st quotient.

2. Divide 1836,88305 by 23.15. Ans. 79.347.
3. Divide .524582192 by .8264. Ans. .63478.
4. Divide .00178600398 by .0463. Ans. .385746.

411	412	413	414	415	416	417	418	419	420
$\frac{1}{411}$	$\frac{1}{412}$	$\frac{1}{413}$	$\frac{1}{414}$	$\frac{1}{415}$	$\frac{1}{416}$	$\frac{1}{417}$	$\frac{1}{418}$	$\frac{1}{419}$	$\frac{1}{420}$

5. Divide $1 by 1.03, and the quotient by 1.03, and this quotient by 1.03, and so on to twenty successive quotients; which reserve as the former.

First quotient, $.97087.

6. Divide $1 by 1.055, and the quotient by 1.055, and this quotient by 1.055, and so on to twenty quotients; which reserve as above. First quotient, $.94786.

7. Divide $1 by 1.045, and the quotient by 1.045, and this quotient by 1.045, and so on to twenty quotients; which reserve as above. First quotient, $.95694.

8. Divide $1 by 1.035, and the quotient by 1.035, and this quotient by 1.035, and so on to twenty successive quotients; which reserve for the purpose of constructing a table of present worths of $1.

First quotient, $.96618.

9. How much per day shall the laborer receive who is to work 359 days for $314.125? *Ans.* $.875.

10. Bought 324 pairs of boots for 607.50 dollars; how much was that per pair? *Ans.* 1$, 87c., 5 m.

11. If 567 yards of calico cost 106$ 31¼ cts., how much was that per yard? *Ans.* $.1875.

12. If 396 yards of silk cost 519 dols. 75 cts., how much was that per yard? *Ans.* 1 dol., 31¼ cts.

13. If 729 barrels of flour cost 5011 dols., 87 cts., 5 m., how much did 1 barrel cost? *Ans.* $6.875.

14. When 112 lbs. of beef cost 9 dols. 80 cts., how much is the cost of 1 lb. *Ans.* 8¾ cts.

15. How much cassimer at $2.645 per yard may be bought for $56.8675? *Ans.* 21.5 yards.

16. At 12½ cents per lb., how much refined sugar can one buy for $90.50? *Ans.* 724 lbs.

17. At 87½ cents per bushel, how much maize will $56.875 purchase? *Ans.* 65 bushels.

18. Laid out in merino sheep $784, the price of each was $3.0625; how many heads did the sum purchase?

Ans. 256 heads.

19. If a grocer should lay out in bacon $182.8125 at the rate of $.1875 for two pounds, how many lbs. of bacon would the sum purchase? *Ans.* 1950 lbs.

421	422	423	424	425	426	427	428	429	430
$\frac{1}{421}$	$\frac{1}{422}$	$\frac{1}{423}$	$\frac{1}{424}$	$\frac{1}{425}$	$\frac{1}{426}$	$\frac{1}{427}$	$\frac{1}{428}$	$\frac{1}{429}$	$\frac{1}{430}$

ADDITION OF MERCANTILE NUMBERS;

OR, ADDITION AND REDUCTION.

The standard of reference for the correction of weights and measures in France is an arc of the meridian. In England it is a pendulum which oscillates seconds in a vacuum, at the level of the sea, in the latitude of London. Such a pendulum is found to measure 39.1+ inches. The cube of one such inch of distilled water, at the temperature of 62° by Fahrenheit's thermometer, being divided into 252458 equal parts, 1000 of these parts will be a troy grain, 7000 of those grains will be a pound avoirdupois, and 5760 of them will be a pound troy. Ten pounds avoirdupois make 1 gallon, which measures 277 cubic inches and $\frac{274}{1000}$ of another inch.

Mercantile Numbers are such as express the divisions of measures of all kinds; viz., those which apply to lines, surfaces, solids, or capacities; force, gravitation, or weight; angles; time.

Rule. Place every denomination under its proper head, or title; then begin at the lowest, and add each column separately, dividing each sum by its tabular number; set down the remainder and carry the quotient to the next higher name, and thus proceed to the last column.

Note. When the tabular number has a cipher, as in shillings, signs, poles, minutes, &c., set down the units' figure of the sum, and divide the rest by the tens only.

Currencies and Sterling Money.

Table or Scale of the Denominations.

4 farthings (marked qr.)	make	1 penny	d.
12 pence	"	1 shilling	s.
20 shillings	"	1 pound	£.

Therefore, qr÷4=d, d÷12=s, s÷20=£.

The farthing is seldom written under its own title, as a

431	432	433	434	435	436	437	438	439	440
$\frac{1}{431}$	$\frac{1}{432}$	$\frac{1}{433}$	$\frac{1}{434}$	$\frac{1}{435}$	$\frac{1}{436}$	$\frac{1}{437}$	$\frac{1}{438}$	$\frac{1}{439}$	$\frac{1}{440}$

number; but generally as a fraction, under the title of pence.

Examples for the Slate.

	£	s.	d.		£	s.	d.		£	s.	d.
1.	1	10	6	2.	39	16	8	3.	34	14	$7\frac{1}{2}$
	6	1	8		24	17	6		28	13	$6\frac{1}{4}$
	7	15	6		43	15	9		12	15	$3\frac{1}{2}$
	2	19	10		91	12	8		69	12	$4\frac{3}{4}$
	3	8	3		87	14	7		56	19	$7\frac{1}{4}$
	11	18	7		65	13	11		75	18	$9\frac{3}{4}$
	£33	14	4		£		?		£		?

4. Set down the following sums in order, and add as above, namely, 1£ 2s. 9d., 2£ 5s. 6d., 3£ 8s. 3d., 4£ 11s., 5£ 13s. 9d.; what is the amount? *Ans.* 17£ 1s. 3d.

5. What is the amount of £1s. $4\frac{1}{2}$d., 1£ 14s. $1\frac{1}{2}$d., 2£ 16s. $10\frac{1}{2}$d., 3£ 19s. $7\frac{1}{2}$d., and 5£ 2s. $4\frac{1}{2}$d.? *Ans.* 14£ 4s. $4\frac{1}{2}$d.

6. How much is the amount of 3£ 8s. $6\frac{3}{4}$d., 4£ 19s. $7\frac{1}{4}$d., 12£ 13s. $8\frac{3}{4}$d., 25£ 14s. $9\frac{1}{4}$d, and 81£ 15s. $8\frac{1}{2}$ d.? *Ans.* 128£ 12s. $4\frac{1}{2}$d.

7. Reduce the following sums, set them down in order, and add them into one sum; viz., 90s., 50s., 130s., 150d., 36d., 220d., 110s., 60s., 70s., 190d., 210d., 70d. *Ans.* 29£ 3s.

8. How much do the following sums amount to, viz., 190d., 110s., 60s., 210d., 70d., and 70s.? *Ans.* 13£ 19s. 2d.

Troy Weight.

Table or Scale of the Denominations.

24 grains, marked gr. = 1 pennyweight, dwt.
20 dwts., or 480 grs. = 1 ounce, oz.
12 oz., or 5760 grs. = 1 pound, lb.

Therefore, grs.÷24=dwts., dwts.÷20=oz., oz.÷12=lbs.

These divisions are used in weighing silver and gold; and by them jewels and liquors are estimated.

441	442	443	444	445	446	447	448	449	450
$\frac{1}{441}$	$\frac{1}{442}$	$\frac{1}{443}$	$\frac{1}{444}$	$\frac{1}{445}$	$\frac{1}{446}$	$\frac{1}{447}$	$\frac{1}{448}$	$\frac{1}{449}$	$\frac{1}{450}$

Examples for the Slate.

	lbs.	oz.	dwts.	grs.		lbs.	oz.	dwts.	grs.
1.	24	11	19	23	2.	35	11	14	18
	12	10	17	21		33	10	13	16
	16	9	15	19		29	8	16	14
	20	8	13	17		31	6	12	15
	32	7	11	15		27	4	18	23
	48	6	9	13		25	2	10	19

3. There are three silver bowls which weigh each 5 lbs. 8 oz. 13 dwts. 18 grs., three dozen spoons which weigh 3 lbs. 7 oz. 9 dwt. 14 grs., and other plate which weighs 20 lbs. 9 oz. 10 dwts. 20 grs. ; what is the collective weight of the whole? *Ans.* 41 lbs. 7 oz. 1 dwt. 16 grs.

4. Five ingots of gold weigh as follows : 9 lbs. 3 oz. 18 dwts. 19 grs., 6 lbs. 3 oz. 12 dwts. 17 grs., 8 lbs. 7 oz. 17 dwts. 21 grs., 7 lbs. 10 oz. 11 dwts. 12 grs., 5 lbs. 14 dwts. 22 grs. ; what is the whole weight of the 5 ingots?
Ans. 37 lbs. 2 oz. 15 dwts. 19 grs.

5. What is the collective weight of ten eagles, each 10 dwts. 18 grs. of gold ; 10 half eagles, each 5 dwts. 9 grs. ; 10 quarter eagles, each 2 dwts. 16½ grs. ; and 10 sovereigns, each 5 dwts. 3 grs.? *Ans.* 11 oz. 19 dwts. 9 grs.

AVOIRDUPOIS WEIGHT.

Table or Scale of the Denominations.

16 drachms, marked dr.=1 ounce, oz.
16 ounces =1 pound, lb.
28 pounds =1 quarter, qr.
4 quarters, or 112 lbs.=1 hundredwt., cwt.
20 cwt., or 2240 lbs. =1 ton, T.

Therefore, dr.÷16=oz., oz.÷16=lbs., lbs.÷28=qrs., qr.÷4, or lbs.÷112=cwts., cwts÷20=T.

Flour, butter, cheese, sugar, coal, iron, &c., are estimated by avoirdupois weight.

451	452	453	454	455	456	457	458	459	460
$\frac{1}{451}$	$\frac{1}{452}$	$\frac{1}{453}$	$\frac{1}{454}$	$\frac{1}{455}$	$\frac{1}{456}$	$\frac{1}{457}$	$\frac{1}{458}$	$\frac{1}{459}$	$\frac{1}{460}$

70 lbs.=1 bushel, English wheat, 8 bushels=1 quarter.

60 lbs.=1 bushel, American wheat, 9⅓ bu.=¼ ton.

56 lbs.=1 bushel Liverpool salt, 4 bu.=1 sack, and 10 sacks=1 ton.

280 lbs.=1 English sack of flour, and 7 sacks=10 American barrels of flour.

64 lbs=1 bushel ground alum salt, and 3½ bu.=1 sack, and 10 sacks=1 ton.

In the case of articles not weighed or measured, names are given to remarkable numbers; thus, 1 is called a piece, per head, &c.; 2, a pair, couple, or brace; 12, a dozen; 20, a score; 12 dozen, or 144, a gross; and 12 gross, a great gross. Of paper, 24 sheets=1 quire, and 20 quires=1 ream. Printers count by the token, of 250 impressions.

Examples for the Slate.

	T.	cwt.	qrs.	lbs
1.	27	14	3	27
	72	13	2	25
	63	12	1	23
	56	18	3	21
	28	9	2	19
	65	15	1	17

	cwt.	qrs.	lbs.	oz.	drs.
2.	31	3	12	15	14
	28	2	14	13	12
	54	1	16	11	10
	63	1	18	9	8
	84	2	22	7	6
	96	3	26	5	4

3. Six barrels of sugar weigh as follows: viz., 8 cwt. 2 qrs. 14 lbs., 9 cwt. 3 qrs. 16 lbs., 7 cwt. 1 qr. 12 lbs., 6 cwt. 3 qrs. 18 lbs., 5 cwt. 24 lbs., and 4 cwt. 3 qrs. 24 lbs.; what is the entire quantity? *Ans.* 42 cwt. 3 qrs. 24 lbs.

4. What quantity of coffee is there in five bags, weighing as follows: viz., 2 qrs. 14 lbs., 3 qrs. 5 lbs., 3 qrs. 17 lbs., 1 cwt. 1 qr. 14 lbs., and 1 cwt. 2 qrs. 27 lbs.
Ans. 5 cwt. 1 qr. 21 lbs.

5. Bought four casks of rice weighing each 2 cwt. 3 qrs. 21 lbs., and three casks weighing each 3 cwt. 19 lbs.; what is the collective weight? *Ans.* 21 cwt. 1 qr. 1 lb.

6. There were seven tubs of butter, four of which

461	462	463	464	465	466	467	468	469	470
$\frac{1}{461}$	$\frac{1}{462}$	$\frac{1}{463}$	$\frac{1}{464}$	$\frac{1}{465}$	$\frac{1}{466}$	$\frac{1}{467}$	$\frac{1}{468}$	$\frac{1}{469}$	$\frac{1}{470}$

weighed thus: 3 cwt. 2 qrs. 12 lbs., 3 cwt. 3 qrs. 16 lbs., 4 cwt. 1 qr. 19 lbs., 2 cwt. 3 qrs. 25 lbs., and each of the other three 5 cwt. 1 qr. 21 lbs.; what was the entire quantity? *Ans.* 31 cwt. 23 lbs.

Apothecaries' Weight.

Table or Scale of the Denominations.

20 grains, marked gr.=1 scruple, ℈ or sc.
3 scruples =1 drachm, ʒ or dr.
8 drachms =1 ounce, ℥ or oz.
12 ounces =1 pound, ℔.

Therefore, gr.÷20=sc., sc.÷3=dr.,
dr.÷8=oz., oz.÷12=lb.

These divisions are used in compounding medicines, but drugs at wholesale are estimated by avoirdupois weight. The lb., oz., gr., are the same as in troy weight.

Examples for the Slate

	℔	oz.	ʒ	℈	gr.		℔	℥	ʒ	℈	gr.
1.	14	11	7	2	18	2.	28	11	1	1	17
	16	9	6	1	16		24	10	2	2	15
	12	8	5	2	14		20	9	3	1	13
	10	7	4	1	12		16	8	4	2	11
	8	6	3	2	10		12	7	5	1	9
	6	5	2	1	8		8	6	6	2	7

3. There are five parcels of drugs, of which the first three contain each 12 lbs. 10 oz. 5 drs. 1 sc. 18 grs., the fourth, 13 lbs. 11 oz. 7 drs. 1 sc. 16 grs., and the fifth, 15 lbs. 8 oz. 6 drs. 1 sc. 16 grs.; what is the collective weight? *Ans.* 68 lbs. 4 oz. 7 drs. 6 grs.

4. What is the total quantity of medicine in the following parcels; namely, 5 lbs. 10 oz. 6 drs. 2 sc. 18 grs., 6 lbs. 9 oz. 5 drs. 1 sc. 16 grs., 7 lbs. 8 oz. 4 drs. 2 ℈ 14 grs., 8 lbs. 6 oz. 2 drs. 1 ℈ 12 grs., 9 lbs. 4 oz. 7 drs. 10 grs., 10 lbs. 11 oz. 5 drs. 17 grs., and 21 lbs. 4 oz. 6 drs. 2 ℈ 19 grs.? *Ans.* 70 lbs. 8 oz. 7 drs. 1 ℈ 6 grs.

471	472	473	474	475	476	477	478	479	480
$\frac{1}{471}$	$\frac{1}{472}$	$\frac{1}{473}$	$\frac{1}{474}$	$\frac{1}{475}$	$\frac{1}{476}$	$\frac{1}{477}$	$\frac{1}{478}$	$\frac{1}{479}$	$\frac{1}{480}$

Long Measure.

Table or Scale of the Denominations.

12 inches, marked in.	=1 foot,	ft.
3 feet, or 36 in.	=1 yard,	yd.
$5\frac{1}{2}$ yards, or $16\frac{1}{2}$ ft.	=1 rod or pole,	po.
40 poles, or 220 yds.	=1 furlong,	fur.
8 furlongs, or 1760 yds.	=1 mile,	m.
3 miles	=1 league,	L.
$69\frac{1}{2}$ statute miles	=1 degree,	deg. or °.
360 degrees	=1 great circle of the earth.	

Therefore, in.÷12=feet, feet÷3=yards, yds.÷$5\frac{1}{2}$=po., po.÷40=furlongs, fur.÷8=m., m.÷$69\frac{1}{2}$=degrees.

In the measure of horses, a hand is 4 inches; of ropes, a fathom is 6 feet. An inch is the least measure that has a name, but parts of an inch are marked upon rules, such as halves, quarters, eighths, tenths, and sixteenths.

Six points are said to make a line, and 12 lines=1 inch, but it is not apparent that such divisions are applied to any practical purpose.

Examples for the Slate.

	deg.	m.	fur.	po.
1.	36	64	7	39
	24	36	6	35
	45	52	5	31
	17	29	4	37
	19	38	3	23
	23	20	2	19

	m.	fur.	po.	yds.
2.	24	7	39	5
	36	1	19	3
	48	6	38	2
	60	2	18	5
	72	5	25	1
	84	3	23	$0\frac{1}{2}$

3. The march of an army for six days was as follows: 25 miles 7 fur. 8 po., 26 miles 3 fur. 27 po., 28 miles 6 fur. 33 po., 30 miles 5 fur. 38 po., 23 miles 5 fur. 29 po. and 31 miles 5 fur. 39 po.; what distance did this army march?

Ans. 167 m. 3 fur. 14 po.

4. The distance from Washington to Dumfries is 37 miles 7 fur. 39 po., thence to Richmond 92 miles 6 fur.

481	482	483	484	485	486	487	488	489	490
$\frac{1}{481}$	$\frac{1}{482}$	$\frac{1}{483}$	$\frac{1}{484}$	$\frac{1}{485}$	$\frac{1}{486}$	$\frac{1}{487}$	$\frac{1}{488}$	$\frac{1}{489}$	$\frac{1}{490}$

38 po., thence to Suffolk 110 m. 1 fur. 3 po., thence to Wilmington, N. C., 240 miles 7 fur. 21 po., thence to Charleston, S. C., 190 miles 19 poles; how much, then, is the distance from Washington to Charleston?

Ans. 672 miles.

5. What is the sum of 4 yds. 2 ft. 11 inches, 5 yards 1 foot 10 inches, 6 yards 1 foot 9 inches, 7 yds. 1 foot 8 inches, 8 yards 2 feet 7 inches, 9 yards 1 foot 6 inches, and 11 yards 3 inches? *Ans.* 54 yards 6 inches.

Cloth Measure.

Table or scale of the Denominations.

$2\frac{1}{4}$ inches, marked in.	=1 nail,	na.
4 nails, or 9 inches	=1 quarter,	qr.
4 quarters, or 16 na.	=1 yard,	yd.
5 quarters, or 45 in.	=1 English ell.	E.E.
3 quarters, or 27 in.	=1 ell Flemish,	E.Fl.

Therefore, inches$\div 2\frac{1}{4}$, or $\frac{4}{9}$ inches=nails, na.$\div$4=qrs., qrs.$\div$4=yds., qrs.$\div$3=E.Fl.

French dry goods are sold by the *aune* and *metre*. The *aune* is about 45 inches, the *metre* 39.37+inches, English statute, which is the standard in the U. States.

The invoices of French goods express the quantity also in yards.

Examples for the Slate.

	yds.	qrs.	na.	in.		E.E.	qrs.	na.	in.
1.	18	3	3	$1\frac{1}{2}$	2.	64	4	3	1
	16	3	1	2		86	3	1	$1\frac{1}{4}$
	36	2	2	$1\frac{3}{4}$		78	2	2	$1\frac{1}{2}$
	42	2	3	1		52	1	3	$1\frac{3}{4}$
	56	1	3	1		45	4	1	2
	64	1	2	$1\frac{1}{4}$		37	3	2	$1\frac{1}{2}$

3. Six pieces of merino plaid measure each 37 yards 3 qrs. 2 na., six pieces measure each 34 yards 2 qrs. 3 na.,

491	492	493	494	495	496	497	498	499	500
$\frac{1}{491}$	$\frac{1}{492}$	$\frac{1}{493}$	$\frac{1}{494}$	$\frac{1}{495}$	$\frac{1}{496}$	$\frac{1}{497}$	$\frac{1}{498}$	$\frac{1}{499}$	$\frac{1}{500}$

six measure each 29 yards 1 qr. 3 na., and six measure each 31 yards 3 qrs. 3 na.; how much did the whole 24 pieces contain?* *Ans.* 803 yds. 2 qrs. 2 na.

4. Eight pieces of calico measure each 33 yds. 2 qrs. 2 na., eight pieces measure each 34 yards 3 qrs. 2 na., eight pieces measure each 35 yards 2 qrs. 3 na., and eight measure each 36 yards 1 qr. 2 na.; how much did the whole 32 pieces measure? *Ans.* 1124 yards 2 qrs.

5. Nine pieces of Holland measure 24 E.Fl. 2 qrs. 3 na. each, nine pieces measure 25 E.Fl. 1 qr. 3 na. each, nine pieces measure 26 E.Fl. 2 qrs. 3 na. each, and nine measure each 27 E.Fl. 1 qr. 2 na.; how much did these 36 pieces measure? *Ans.* 944 E.Fl. 3 na.

Square Measure.

Table or Scale of the Denominations.

144 square inches, sq. in.	=1 square foot,	sq. ft.
9 square feet	=1 square yard,	sq. yd.
$30\frac{1}{4}$ sq. yards or $272\frac{1}{4}$ sq. ft.	=1 square rod,	sq. po.
40 sq. poles, or 1210 sq. yds.	=1 rood,	R.
4 roods, or 160 sq. po.	=1 acre,	A.

Therefore, sq. in.÷144=sq. ft., sq. ft.÷9=sq. yds., sq. yds.÷$30\frac{1}{4}$=sq. po., sq. po.÷40=roods, roods÷40=acres.

This measure is used in all superficies, where length and breadth are to be considered, as in farms or lots; also in boards, glass, pavements, plastering, wainscoting, roofing, flooring, &c.

But artificers have another measure, called the duodecimal, in which 12 fourths ''''=1 third ''', 12 thirds=1 second '', 12 seconds =1 inch, 12 inches=1 foot.

Surveyors use Gunter's chain of 4 poles, or 66 feet=792 inches; this chain is divided into 100 links, each link being 7.92 inches. A square chain=16 square poles, and 10 square chains make 160 sq. po.=1 acre.

---

* Add the yds. in one piece of each parcel, and multiply the *sum* by the number of pieces in *one* parcel, and so in all similar cases; because the sum of the products is equal to the product of the sum. Or thus: the rectangle of two numbers is equal to the several rectangles of one of them and all the parts of the other.

501	502	503	504	505	506	507	508	509	510
$\frac{1}{501}$	$\frac{1}{502}$	$\frac{1}{503}$	$\frac{1}{504}$	$\frac{1}{505}$	$\frac{1}{506}$	$\frac{1}{507}$	$\frac{1}{508}$	$\frac{1}{509}$	$\frac{1}{510}$

A square is a figure whose length is equal to its breadth, and whose angles contain 90° each, or $\frac{1}{4}$ the compass of the point in which the lines meet.

Examples for the Slate.

	A.	R.	po.	yds.		A.	R.	po.	yds.
1.	24	3	39	30	2.	75	1	20	20
	34	2	36	28		65	2	24	18
	32	1	33	26		55	3	28	22
	28	3	30	37		45	1	32	16
	56	2	27	28		35	2	36	24
	48	1	24	32½		25	3	18	21

3. There are four fields which measure each 18 A. 2 R. 20 po., four which measure each 16 A. 3 R. 36 po., four which measure each 18 A. 2 R. 28 po., four which measure each 22 A. 1 R. 24 po., and four which measure each 24 A. 2 R. 29 po.; how much land do the twenty fields contain? *Ans.* 405 A. 1 R. 28 po.

4. The flooring of 6 rooms is 15 yards 6 ft. 48 in. each, of other six each 16 yds. 6 ft. 36 in., of other six, 17 yds. 7 ft. 72 in. each, of other six 18 yards 5 ft. 24 in. each, of other six, 19 yds. 3 ft. 108 in. each, and of other six, 20 yds. 4 ft. 96 in. each; how much is the whole quantity in the 36 rooms? *Ans.* 652 yds. 4 feet.

5. A certain house has twelve windows which measure each 42 ft. 72 in., and twelve which measure each 38 ft. 84 in., and twelve which measure each 34 ft. 96 in., and twelve which measure each 30 ft. 120 in., and twelve which measure each 26 ft. 48 in. Now it is required to know the quantity of glazing in the 60 windows?

Ans. 2075 feet.

6. The plastering of five rooms measures each 71 yds. 6 ft. 108 in., of other five, each 66 yds. 8 ft. 96 in., of other five, each 61 yds. 4 ft. 84 in., of other five, each 56 yds. 5 ft. 72 in., of other five, each 51 yds. 7 ft. 60 in.; what quantity of plastering does these 25 rooms measure?

Ans. 1543 yds. 2 ft. 84 in.

511	512	513	514	515	516	517	518	519	520
$\frac{1}{511}$	$\frac{1}{512}$	$\frac{1}{513}$	$\frac{1}{514}$	$\frac{1}{515}$	$\frac{1}{516}$	$\frac{1}{517}$	$\frac{1}{518}$	$\frac{1}{519}$	$\frac{1}{520}$

Liquid Measure.

Table or Scale of the Denominations.

4 gills, marked gls.	=1 pint,	pt.
2 pints	=1 quart,	qt.
4 quarts	=1 gallon,	gal.
$31\frac{1}{2}$ gallons	=1 barrel,	bl.
42 gallons	=1 tierce,	tier.
63 gals., or 2 bls.	=1 hogshead,	hhd.
84 gals., or 2 tier.	=1 puncheon,	pun.
126 gals., or 2 hhds.	=1 pipe or butt,	pi.
252 gals., or 2 pipes	=1 tun,	T.

Therefore, gls.÷4=pints, pints÷8=gals., gals.÷$31\frac{1}{2}$ bls., gals.÷42=tier., gals.÷63=hhds., gals.÷84=pun.

The gallon is the unit of this measure; the capacities of larger vessels are estimated by the gallon.

This measure is used for ale, beer, wines, spirits, cider, mead, perry, vinegar, oil, molasses, honey, &c. But the gallons for wine, ale, and dry measures, are not of the same capacity, as shall be noticed under the title of cubic or solid measure.

Examples for the Slate.

	hhds.	gals.	qts.		gals.	qts.	pts.	gls.
1.	29	62	3	2.	55	3	1	1
	32	60	2		49	2	1	2
	35	58	1		43	1	1	3
	38	56	3		37	1	1	3
	41	54	2		31	2	1	2
	44	52	1		25	3	1	1

3. A distiller had 6 puncheons of molasses, containing each 82 gals. 3 qts., 6 puncheons containing each 83 gals. 3 qts., 6 puncheons containing each 85 gals. 2 qts., 6 puncheons containing each 78 gals. 1 qt., and 6 puncheons containing each 80 gals. 2 qts.; how many gallons did the 30 puncheons contain? *Ans.* 2464 gals. 2 qts.

521	522	523	524	525	526	527	528	529	530
$\frac{1}{521}$	$\frac{1}{522}$	$\frac{1}{523}$	$\frac{1}{524}$	$\frac{1}{525}$	$\frac{1}{526}$	$\frac{1}{527}$	$\frac{1}{528}$	$\frac{1}{529}$	$\frac{1}{530}$

4. There are two barrels of cider which contain each 25 gals. 3 qts. 1 pt., two which contain each 26 gals. 2 qts. 1 pt., two which contain each 27 gals. 1 qt. 1 pt., two, each 28 gals. 3 qts., two, each 29 gals. 2 qts. 1 pt., two, each 30 gals. 1 qt., two, each 31 gals. 3 qts. 1 pt., and two, each 32 gals. 3 qts. 1 pt.; what quantity do the 16 barrels contain? *Ans.* 466 gals. 2 qts.

5. A whale ship had on board seven vessels containing each 76 gals. 3 qts. of sperm oil, seven containing each 84 gals. 3 qts., seven containing each 82 gals. 2 qts., seven containing each 56 gals. 3 qts., seven containing each 48 gals. 1 qt., seven containing each 54 gals. 3 qts.; how much oil did the 42 vessels contain?

Ans. 2826 gals. 1 quart.

Dry Measure.

Table or Scale of the Denominations.

2 pints, marked pts.=1 quart, qt.
8 quarts, or 16 pts. =1 peck, pe.
4 pecks, or 32 qts. =1 bushel, bu.

Therefore, pts.÷2=qts., quarts÷8=pecks, pecks÷4, or qts.÷32=bushels.

This measure is used for grains, seeds, roots, fruit, salt, coal, lime, oysters, sand, &c. See under the title of Avoirdupois Weight, the names of other dry measures; and under the head of cubic measure the solid contents of the bushel.

Examples for the Slate.

	bu.	pe.	qts.	pts.		bu.	pe.	qts.	pts.
1.	72	3	7	1	2.	28	1	2	1
	68	2	6	1		26	3	4	1
	64	1	5	1		82	2	6	1
	60	3	4	1		62	1	7	1
	56	2	3	1		68	3	5	1
	52	1	2	1		86	2	3	1

531	532	533	534	535	536	537	538	539	540
$\frac{1}{531}$	$\frac{1}{532}$	$\frac{1}{533}$	$\frac{1}{534}$	$\frac{1}{535}$	$\frac{1}{536}$	$\frac{1}{537}$	$\frac{1}{538}$	$\frac{1}{539}$	$\frac{1}{540}$

3. A farmer brought to town 68 bu. 3 pe. 7 qts. of oats, 75 bu. 2 pe. 6 qts. of wheat, 250 bu. 2 pe. 5 qts. of corn, 96 bu. 1 pe. 3 qts. of rye, 136 bu. 2 pe. 4 qts. of potatoes, 7 bu. 3 pe. 6 qts. of clover seed, 6 bu. 1 pe. 3 qts. of timothy seed, 27 bu. 6 qts. of flax seed, 84 bu. 2 pe. 2 qts. of turnips, 56 bu. 2 pe. 6 qts. of apples, and 350 bushels of peaches; how many bushels did this useful citizen bring in? *Ans.* 1161 bu.

4. A merchant bought four loads of salt, each 78 bu. 3 pe. 6 qts., and four loads, each 87 bu. 2 pe. 4 qts., and four loads, each 84 bu. 3 pe. 5 qts., and four loads, each 92 bu. 1 pe. 7 qts., and four loads, each 69 bu. 2 pe. 3 qts.; how many bushels did the twenty loads contain? *Ans.* 1654 bu. 4 qts.

Time.

Table or Scale of the Denominations.

60 seconds, marked sec., or ″	=1 minute,	m. or ′.
60 minutes	=1 hour,	hr.
24 hours	=1 day,	d.
7 days	=1 week,	w.
28 days	=1 lunar month,	l. m.,
28, 29, 30, or 31 days	=1 calendar mo.,	c. m.
12 calendar or 13 lunar months	=1 year,	yr.

Therefore, sec.÷60=m., m.÷60=hrs., hrs.÷24=ds., ds.÷7=ws., ds.÷365¼=years.

The earth is its own chronometer; its rotation on its axis forms the alternate change of light and shade; its revolution in its orbit forms the seasons and the year. But clocks, watches, dials, and sand-glasses mark the smaller divisions.

Instead of 365¼ days the common years consist of 365 days, and every fourth year consists of 366 days, the day formed of the four quarters of a day being added to February in that year. Every fourth year is therefore called leap year or bissextile; yet not *every* fourth year, because in 400 years there are only 97 leap years; the remaining 303 are common years. The 4 years of John Adams' presidential term were shorter by 1 day than those of any other of the Presidents, the centennial year being a common one.

The number of days in each of the calendar months in common years may be seen as follows: viz.,

541	542	543	544	545	546	547	548	549	550
$\frac{1}{541}$	$\frac{1}{542}$	$\frac{1}{543}$	$\frac{1}{544}$	$\frac{1}{545}$	$\frac{1}{546}$	$\frac{1}{547}$	$\frac{1}{548}$	$\frac{1}{549}$	$\frac{1}{550}$

1. January, 31.	5. May, 31.	9. September, 30.
2. February, 28.	6. June, 30.	10. October, 31.
3. March, 31.	7. July, 31.	11. November, 30.
4. April, 30.	8. August, 31.	12. December, 31.

The intercalary day is added to the second month, in leap years.

Examples for the Slate.

	yrs.	mos.	ws.	ds.		ds.	hrs.	m.	sec.
1.	35	11	3	6	2.	29	23	56	55
	33	9	2	1		31	21	52	51
	31	7	1	5		33	19	48	47
	29	5	2	2		35	17	44	43
	27	3	3	4		37	15	40	39
	26	1	1	3		39	13	36	35

3. What is the sum of 25 years 10 mos. 26 ds., 29 years 7 mos. 21 ds., 34 years 8 mo. 18 days, 58 years 11 mo. 24 ds., 24 yrs. 6 mos. 16 ds., and 33 yrs. 9 mos. 28 ds. ?
Ans. 207 yrs. 7 mos. 13 ds.

Note. In adding yrs. mo. ds., the divisors will be 30 and 12; but for yrs. mos. ws. ds. the divisors will be 7, 4, and 13.

To determine the number of days from one date to another, the latter days of the former month + the former days of the latter month + the days of the intervening months, if any, will be the number.

4. What day of the year was the 4th of July, 1847 ?
Ans. 185th day.

5. From the 7th of the 2d month to the 17th of the 10th month, 1847, inclusive, how many days ?
Ans. 253 days.

Angular or Circle Measure.

Table or Scale of the Denominations.

60 seconds sec., or ″	=1 minute,	m. or ′.
60 minutes	=1 degree,	deg. or °.
30 degrees	=1 sign,	sig.
90 degrees	=1 quadrant,	qdt.
12 signs, or 360°	=1 circle.	

Therefore, sec.÷60=m., m.÷60=degrees, degs.÷30=sigs., sigs.÷12=circle, qdts.÷4=the circle, or revolution.

551	552	553	554	555	556	557	558	559	560
$\frac{1}{551}$	$\frac{1}{552}$	$\frac{1}{553}$	$\frac{1}{554}$	$\frac{1}{555}$	$\frac{1}{556}$	$\frac{1}{557}$	$\frac{1}{558}$	$\frac{1}{559}$	$\frac{1}{560}$

This measure is used in geography, navigation, and astronomy; it is also used in surveying, and in all the works of art in which the heights or distances of objects are inaccessible.

Every circle, great or small, is supposed to be divided into 360°, and the number of degrees between two distant objects is ascertained by the angle formed by the lines leading from the two objects to their point of meeting: thus, if distant objects are seen at A, C, or G, H (see the figure in the margin), the observer at B can determine, by his instruments, the number of degrees intercepted between them.

The signs are twelve, extending around the celestial globe, forming the zodiac, and about 8° on either side of the ecliptic.

Examples for the Slate.

	sigs.	°	′	″
1.	1	29	59	54
	1	28	58	53
	1	27	57	52
	1	26	56	51
	2	25	55	50

	sigs.	°	′	″
2.	3	24	40	44
	2	23	48	43
	1	22	47	42
	1	21	46	41
	2	20	45	40

Note. Differences of latitude and longitude, when the places are in opposite hemispheres, are found by addition; when in the same hemisphere, by subtraction.

3. Philadelphia lies in longitude 75° 14′ west from London, and Constantinople in longitude 28° 55¼′ east; what is their difference of longitude? *Ans.* 104° 9¼′

4. Constantinople lies in latitude 41° 1′ north, and Cape Horn in 55° 59′ south; what is their difference of latitude? *Ans.* 97 degrees.

5. Philadelphia lies in latitude 39° 56½′ north, and the Cape of Good Hope in latitude 34° 29′ south; what is their difference of latitude? *Ans.* 74° 25½′.

6. Canton in China lies in longitude 113° 16′ east, and Funchal in Madeira in longitude 17° 5′ west; what is their difference of longitude? *Ans.* 130° 21′.

7. The latitude of the north pole is 90°, and of the south pole 90°; what is their difference of latitude?

561	562	563	564	565	566	567	568	569	570
$\frac{1}{561}$	$\frac{1}{562}$	$\frac{1}{563}$	$\frac{1}{564}$	$\frac{1}{565}$	$\frac{1}{566}$	$\frac{1}{567}$	$\frac{1}{568}$	$\frac{1}{569}$	$\frac{1}{570}$

Cubic or Solid Measure.

A cube has six square and equal sides (see the figure in the margin).

It is the measuring unit for all solid and capacious bodies.

The yard is the standard unit of measure: all other measures are referred for adjustment to the yard, as multiples or parts of it. The yard itself refers to the pendulum; the pendulum to the law of gravitation, affected by the latitude and the spheroidal figure of the earth.

A cubic yard	=	27 cubic feet.
A cubic foot	=	1728 cubic inches.
An ale gallon	=	282 cubic inches.
A wine gallon	=	231 cubic inches.
A dry gallon	=	268.8 cubic inches.
A bushel, 8 gallons	=	2150.4 cubic inches.
A cord of wood	=	128 cubic feet.
A ton of round timber	=	40 cubic feet.
A ton of hewn timber	=	50 cubic feet.

The tons of round and hewn timber refer to the equal spaces which they occupy in the hold of a ship.

SUBTRACTION OF MERCANTILE NUMBERS;

OR, SUBTRACTION AND REDUCTION.

When any quantity loses its half, it has $\frac{1}{2}$ left. If it loses $\frac{1}{3}$, it has $\frac{2}{3}$ left. Take away $\frac{1}{4}$, it has $\frac{3}{4}$ left. If $\frac{1}{5}$ be subtracted, $\frac{4}{5}$ remain; and so on.

This arises from the fact that every quantity has two halves, three thirds, four fourths, five fifths, &c.

Now a part so taken is called an *aliquot* part, and the

571	572	573	574	575	576	577	578	579	580
$\frac{1}{571}$	$\frac{1}{572}$	$\frac{1}{573}$	$\frac{1}{574}$	$\frac{1}{575}$	$\frac{1}{576}$	$\frac{1}{577}$	$\frac{1}{578}$	$\frac{1}{579}$	$\frac{1}{580}$

parts left are called the *complement* of such an aliquot or single part.

The class will join in the following exercise:

1. If from a dollar we take $\frac{1}{2}$ of it, how many cents are left? If $\frac{1}{3}$ of the dollar be taken, how many cents are left? If $\frac{1}{4}$ of it be taken, how many cents are left? And so, having the $ changed to cents,

If $\frac{1}{5}$, how many? If $\frac{1}{12}$, how many?
If $\frac{1}{6}$, how many? If $\frac{1}{16}$, how many?
If $\frac{1}{8}$, how many? If $\frac{1}{20}$, how many?
If $\frac{1}{10}$, how many? If $\frac{1}{25}$, how many?

2. If from a £ currency its half be taken, how many shillings are left? If $\frac{1}{3}$ of the £ be taken, how much is left? If $\frac{1}{4}$ how much? And so, the £ being reduced to shillings and pence,

If $\frac{1}{5}$, how much left? If $\frac{1}{10}$, how much?
If $\frac{1}{6}$, how much? If $\frac{1}{12}$, how much?
If $\frac{1}{8}$, how much? If $\frac{1}{20}$, how much?

3. If from 1 lb. troy its half be taken, how many ounces are left? If $\frac{1}{3}$ of the lb. be taken, how many ounces are left? And so, having the lb. reduced to oz., dwts., or grs.,

If $\frac{1}{4}$, how many left? If $\frac{1}{10}$, how many?
If $\frac{1}{5}$, how many? If $\frac{1}{12}$, how many?
If $\frac{1}{6}$, how many? If $\frac{1}{16}$, how many?
If $\frac{1}{8}$, how many? If $\frac{1}{20}$, how many?

4. If from 1 ton, or 20 cwt., its half be taken, how many cwt. are left? If $\frac{1}{4}$ of the ton be taken, how much is left? And so, the ton being reduced to cwt., qrs., or lbs.,

If $\frac{1}{5}$, how much left? If $\frac{1}{14}$, how much?
If $\frac{1}{7}$, how much? If $\frac{1}{16}$, how much?
If $\frac{1}{8}$, how much? If $\frac{1}{20}$, how much?
If $\frac{1}{10}$, how much? If $\frac{1}{28}$, how much?
If $\frac{1}{112}$, how much?

5. If from one lb. apothecaries' weight, its half be taken, how many ounces are left? If $\frac{1}{3}$ of the lb. be

581	582	583	584	585	586	587	588	589	590
$\frac{1}{581}$	$\frac{1}{582}$	$\frac{1}{583}$	$\frac{1}{584}$	$\frac{1}{585}$	$\frac{1}{586}$	$\frac{1}{587}$	$\frac{1}{588}$	$\frac{1}{589}$	$\frac{1}{590}$

taken, how many ounces are left? And so, the lb. being reduced to oz., drs., sc., or grs., if $\frac{1}{4}$ be taken, how much is left?

If $\frac{1}{5}$, how much? If $\frac{1}{16}$, how much?
If $\frac{1}{6}$, how much? If $\frac{1}{18}$, how much?
If $\frac{1}{8}$, how much? If $\frac{1}{20}$, how much?
If $\frac{1}{12}$, how much? If $\frac{1}{24}$, how much?

6. If from 1 mile its half be taken, how much is left? If $\frac{1}{3}$ of the mile be taken, how much is left? And so, the mile being reduced to fur., po., yds., or ft.,

If $\frac{1}{4}$, how much? If $\frac{1}{16}$, how much?
If $\frac{1}{5}$, how much? If $\frac{1}{20}$, how much?
If $\frac{1}{8}$, how much? If $\frac{1}{24}$, how much?
If $\frac{1}{10}$, how much? If $\frac{1}{32}$, how much?

7. If from 1 yard its half be taken, how many quarters are left? If $\frac{1}{3}$ be taken, how many inches are left? And so, the yard being reduced to qrs., na., inches,

If $\frac{1}{4}$, how much? If $\frac{1}{9}$, how much?
If $\frac{1}{6}$, how much? If $\frac{1}{12}$, how much?
If $\frac{1}{8}$, how much? If $\frac{1}{16}$, how much?

8. If from 1 square yard its half be taken, how many feet and inches are left? If $\frac{1}{3}$ of the yard be taken, how much is left? And so, the yard being reduced to feet and inches,

If $\frac{1}{4}$, how many? If $\frac{1}{16}$, how many?
If $\frac{1}{6}$, how many? If $\frac{1}{18}$, how many?
If $\frac{1}{8}$, how many? If $\frac{1}{24}$, how many?
If $\frac{1}{12}$, how many? If $\frac{1}{36}$, how many?

If from one acre its half be taken, how many roods are left? If the acre lose $\frac{1}{4}$, how much is left? And so, the acre being reduced to sq. po., sq. yds.,

If $\frac{1}{8}$, how much? If $\frac{1}{32}$, how much?
If $\frac{1}{10}$, how much? If $\frac{1}{40}$, how much?
If $\frac{1}{16}$, how much? If $\frac{1}{60}$, how much?
If $\frac{1}{18}$, how much? If $\frac{1}{80}$, how much?

9. If from one gallon its half be taken, how much is left? If $\frac{1}{4}$ of the gallon be taken, how much is left? If $\frac{1}{8}$, how much? If $\frac{1}{16}$, how much?

591	592	593	594	595	596	597	598	599	600
$\frac{1}{591}$	$\frac{1}{592}$	$\frac{1}{593}$	$\frac{1}{594}$	$\frac{1}{595}$	$\frac{1}{596}$	$\frac{1}{597}$	$\frac{1}{598}$	$\frac{1}{599}$	$\frac{1}{600}$

10. If from 1 bushel its half be taken, how much is left? If $\frac{1}{4}$, how much? If $\frac{1}{8}$, how much? If $\frac{1}{16}$, how much?

11. If from 1 year its half be taken, how many months are left? If from 1 leap year $\frac{1}{3}$ of the days be taken, how many are left? If $\frac{1}{4}$ of its weeks be taken from a year, how many are left? If $\frac{1}{5}$ of its days be taken from a common year, how many are left? If $\frac{1}{6}$ of the calendar months be taken from a year, how many are left? If $\frac{1}{12}$, how many?

12. If from the circle of the zodiac half the signs be taken, how many will be left? If from any circle $\frac{1}{4}$ or a quadrant be taken, how many degrees are left? If $\frac{1}{5}$ be taken, how many left? If $\frac{1}{6}$, how many? If $\frac{1}{8}$, how many? If $\frac{1}{9}$, how many? If $\frac{1}{10}$, how many? If $\frac{1}{12}$, how many? If $\frac{1}{18}$, how many? If $\frac{1}{24}$, how many?

13. If from 1 cubic yard $\frac{1}{3}$ of it be taken, how many cubic feet are left? If $\frac{1}{9}$, how many? If $\frac{1}{27}$, how many?

If from a cubic foot its half be taken, how many cubic inches are left? If $\frac{1}{3}$, how many? If $\frac{1}{4}$, how many? If $\frac{1}{6}$, how many? If $\frac{1}{8}$, how many? If $\frac{1}{9}$, how many? If $\frac{1}{12}$, how many? If $\frac{1}{16}$, how many? If $\frac{1}{18}$, how many?

If from one cord of wood its half be taken, how many cubic feet are left? If $\frac{1}{4}$, how many? If $\frac{1}{8}$, how many? If $\frac{1}{16}$, how many? If $\frac{1}{32}$, how many? If $\frac{1}{64}$, how many? If $\frac{1}{128}$, how many?

Rule. Place the less quantity under the greater, so that units of equal value be in the same column; then, beginning at the right, take the lower number from the upper, if possible; if not, take it from the unit of the next higher name reduced,* and to the remainder add the upper number; set down the sum, and add 1 to the next lower place for the reduced unit, if any, by which the upper was increased, and so proceed to the end.

* This is to increase the upper number by a unit of a certain value; but if the minuend be increased, it is proper also to increase the subtrahend equally; and if there were 2, 3, or more units added to each of the given terms, it would in no wise affect the difference.

601	602	603	604	605	606	607	608	609	610
$\frac{1}{601}$	$\frac{1}{602}$	$\frac{1}{603}$	$\frac{1}{604}$	$\frac{1}{605}$	$\frac{1}{606}$	$\frac{1}{607}$	$\frac{1}{608}$	$\frac{1}{609}$	$\frac{1}{610}$

Examples of Currencies and Sterling Money.

	£	s.	d.		£	s.	d.		£	s.	d.
1.	34	16	8	2.	31	11	$5\frac{1}{2}$	3.	25	18	$1\frac{1}{4}$
	17	14	6		13	16	$8\frac{1}{4}$		18	18	$10\frac{1}{2}$

4. Borrowed £357 16s. 8d., and paid £186 17s. 9d.; what remains due? *Ans.* £170 18s. 11d.

5. What is the difference between £556 14s. $8\frac{1}{4}$d. and £328 17s. 9d.? *Ans.* £227 16s. $11\frac{1}{4}$d.

6. Lent £3 11s. $4\frac{1}{2}$d. and received £2 16s. $10\frac{1}{2}$d.; what is yet unpaid? *Ans.* 14s. 6d.

7. Having borrowed £100, and paid at one time £39 15s. $8\frac{1}{4}$d., at another payment £43 7s. $5\frac{1}{2}$d., what do I yet owe? *Ans.* £16 16s. $10\frac{1}{4}$d.

8. A bill of £150 is paid as follows, viz.: A's note is given for £15 5s. B's do. for £10, an order for £35 7s. 7d. and bank notes for £80; how much is the balance? *Ans.* £9 7s. 5d.

Examples of Troy Weight.

	lbs.	oz.	dwts.	grs.		lbs.	oz.	dwts.	grs.
1.	39	10	18	21	2.	35	11	12	14
	19	11	17	22		18	10	15	18

3. When 14 lbs. 11 oz. 18 dwts. 21 grs. is deducted from 28 lbs. 10 oz. 12 dwts. 22 grs., what remains? *Ans.* 13 lbs. 10 oz. 14 dwts. 1 gr.

4. A silversmith bought 7 lbs. 5 oz. 14 dwts. 21 grs. of old silver, of which he made 3 lbs. 9 oz. 18 dwts. 18 grs. into spoons; what quantity remains unwrought? *Ans.* 3 lbs. 7 oz. 16 dwts. 3 grs.

5. From 6 old eagles (11 dwts. 6 grs. each) having melted 1 oz. 8 dwts. 16 grs. of gold, what quantity doth yet remain? *Ans.* 1 oz. 18 dwts. 20 grs.

6. From a mass of gold weighing 3 oz. 2 dwts. 9 grs. having taken 1 oz. 5 dwts. 18 grs., what is the remainder? *Ans.* 1 oz. 16 dwts. 15 grs.

7. A silversmith fuses 4 French crowns of 19 dwts. each, and 4 dollars of 17 dwts. 8 grs. each, and from the

611	612	613	614	615	616	617	618	619	620
$\frac{1}{611}$	$\frac{1}{612}$	$\frac{1}{613}$	$\frac{1}{614}$	$\frac{1}{615}$	$\frac{1}{616}$	$\frac{1}{617}$	$\frac{1}{618}$	$\frac{1}{619}$	$\frac{1}{620}$

mass takes 4 oz. 8 dwts. 20 grs. ; how much does yet remain ? *Ans.* 2 oz. 16 dwts. 12 grs.

Examples of Avoirdupois Weight.

	T.	cwt.	qrs.	lbs.	oz.		cwt.	qrs.	lbs.	oz.	drs.
1.	61	9	2	14	12	2.	84	3	17	10	12
	16	16	3	22	13		48	2	20	14	14

3. Bought 14 cwt. 2 qrs. 16 lbs. of sugar, and sold 9 cwt. 3 qrs. 27 lbs. ; what quantity is left? *Ans.* 4 cwt. 2 qrs. 17 lbs.

4. Three parcels, viz.: 3 cwt. 1 qr. 14 lbs., 2 cwt. 1 qr. 13 lbs., and 3 cwt. 15 lbs., being taken from a hhd. containing 18 cwt. 1 qr. 24 lbs., query the remainder? *Ans.* 9 cwt. 2 qrs. 10 lbs.

5. From a hhd. containing 10 cwt. 2 qrs. 16 lbs. of rice, having sold 3 cwt. 23 lbs. how much is left? *Ans.* 7 cwt. 1 qr. 21 lbs.

6. What is the difference of weight between two bales of cotton of 3 cwt. 15 lbs., and 7 cwt. 2 qrs. 2 lbs.? *Ans.* 4 cwt. 1 qr. 15 lbs.

7. What is the difference between 1 cwt. 2 qrs. 15 lbs. +1 cwt. 3 qrs. 5 lbs.+1 cwt. 3 qrs. 25 lbs. and 3 qrs. 17 lbs.+3 qrs. 25 lbs.+1 cwt. 1 qr. 17 lbs.? *Ans.* 2 cwt. 14 lbs.

Examples of Apothecaries' Weight.

	lbs.	oz.	drs.	sc.	grs.		℔	℥	ʒ	℈	grs.
1.	96	8	6	1	16	2.	32	6	4	2	12
	69	10	7	2	18		23	8	6	1	14

3. How much is 18 lbs. 9 oz. 7 drs. 2 sc. 14 grs. greater than two parcels, each 6 lbs. 11 oz. 4 drs. 1 sc. 9 grs.? *Ans.* 4 lbs. 10 oz. 6 drs. 2 sc. 16 grs.

4. Two parcels of drugs weigh each 5 lbs. 4 oz. 6 drs. 2 sc. 12 grs., and two others weighing 8 lbs. 4 drs. 14 grs., and 7 lbs. 9 oz. 2 sc. ; what is their difference of weight? *Ans.* 4 lbs. 11 oz. 7 drs. 10 grs.

5. Take 5 lbs. 11 oz. 7 drs. 2 sc. 18 grs. from 6 lbs. 5 oz. 3 drs. 1 sc. 4 grs. *Ans.* 5 oz. 3 drs. 1 sc. 6 grs.

621	622	623	624	625	626	627	628	629	630
$\frac{1}{621}$	$\frac{1}{622}$	$\frac{1}{623}$	$\frac{1}{624}$	$\frac{1}{625}$	$\frac{1}{626}$	$\frac{1}{627}$	$\frac{1}{628}$	$\frac{1}{629}$	$\frac{1}{630}$

6. If a compound of medicine weigh 11 lbs. 9 oz. 6 drs. 2 sc. 18 grs., and 10 lbs. 8 oz. 7 drs. 1 sc. 19 grs. thereof be sold, how much remains? *Ans.* 1 lb. 7 drs. 19 grs.

7. The less quantity is 5 lbs. 10 oz. 6 drs. 2 grs., the greater 16 lbs. 9 oz. 7 drs. 2 sc. 1 gr., what is their difference? *Ans.* 10 lbs. 11 oz. 1 dr. 1 sc. 19 grs.

Examples of Long Measure.

	deg.	m.	fur.	po.		yds.	ft.	in.
1.	42	26	4	29	2.	31	1	8
	24	56	6	32		13	2	6

3. From 5 yds. 2 ft. $9\frac{2}{3}$ in., take 2 pieces each 2 yds. $9\frac{2}{3}$ in.; what remains? *Ans.* 1 yd. 1 ft. $2\frac{1}{3}$ in.

4. If 36 m. 5 fur. 30 po. of road is to be made by two men, whereof one has 21 m. 7 fur. 33 po., what is left for the other? *Ans.* 14 m. 5 fur. 37 po.

5. If two men start for a place and travel four days, one of them to go 28 m. 5 fur. 37 po. per day, the other 22 m. 6 fur. 28 po. per day; how far has the former outgone the latter? *Ans.* 23 m. 4 fur. 36 po.

6. What is the difference in the length of two cables, one of which is 108 fath. 4 ft. 8 in., the other 91 fath. 5 ft. 9 in.? *Ans.* 16 fath. 4 ft. 11 in.

7. If Pittsburg lie 298 m. 4 fur. 29 po. from Philadelphia, and 300 m. 3 fur. 16 po. from Washington, what is the difference in the length? *Ans.* 1 m. 6 fur. 27 po.

Examples of Cloth Measure.

	yds.	qrs.	na.		E.E.	qrs.	na.		E.Fl.	qrs.	na.
1.	42	2	1	2.	31	2	2	3.	41	1	3
	24	3	2		16	4	3		14	2	1

4. From 200 E.E. take 156 E.E. 4 qrs. 3 na., and what remains? *Ans.* 43 E.E. 1 na.

5. From 200 yards take 156 yds. 3 qrs. 3 na.; how much is left? *Ans.* 43 yds. 1 na.

6. From 200 E.Fl. take 156 E.Fl. 2 qrs. 3 na., and how much remains? *Ans.* 43 E.Fl. 1 na.

631	632	633	634	635	636	637	638	639	640
$\frac{1}{631}$	$\frac{1}{632}$	$\frac{1}{633}$	$\frac{1}{634}$	$\frac{1}{635}$	$\frac{1}{636}$	$\frac{1}{637}$	$\frac{1}{638}$	$\frac{1}{639}$	$\frac{1}{640}$

7. How much are 4 webs of 25 yds. 3 qrs. 3 na. each greater than 5 webs of 17 yds. 2 qrs. 2 na. each?

Ans. 15 yds. 2 qrs. 2 na.

Examples of Square Measure.

	A.	R.	po.	yds.		A.	R.	po.		A.	R.	po.	yds.
1.	56	2	32	24	2.	34	1	26	3.	62	1	24	28
	28	3	29	28¼		17	0	33		26	3	35	19¼

4. How much is 43 A. 1 R. 24 po. 16 yds. greater than 34 A. 2 R. 26 po. 28 yds.?

Ans. 8 A. 2 R. 37 po. 18¼ yds.

5. The floor of a large room is 24 yds. 5 ft. 108 in. in content, whereof a part containing 10 yds. 8 ft. 132 in. is walled off. Query, the remainder?

Ans. 13 yds. 5 ft. 120 in.

6. The plastering of a house measures 426 yds. 5 ft. 72 in., and is done by two workmen; the part of one is 218 yds. 8 ft. 84 in.; what is the share of the other?

Ans. 207 yds. 5ft. 132 in.

7. The quantity of glazing in the front of a building is 174 ft. 36 in., whereof the lower story has 78 ft. 108 in.; what remains for the others? *Ans.* 95 ft. 72 in.

8. When Mephibosheth divided the land with his servant, of 306 A. 2 R. 28 po. 21 yds., he retained 158 A. 3 R. 32 po. 28 yds.; how much had Ziba?

Ans. 147 A. 2 R. 35 po. 23¼ yds.

Examples of Liquid Measure.

	T.	hhds.	gals.	qts.		hhds.	gals.	qts.	pts.
1.	191	2	42	2	2.	125	24	1	1
	69	2	56	3		48	42	3	1

3. What is the difference between 55 T. 1 hhd. 3 qts., and 49 T. 58 gals. 2 qts.? *Ans.* 6 T. 5 gals. 1 qt.

4. What remains when 4 hhds. 48 gals. 2 qts. 1 pt. is drawn from a large stock cask containing 9 T. 2 hhds. 1 pt.? *Ans.* 8 T. 1 hhd. 14 gals. 2 qts.

5. The difference of the capacity of two vessels is 35

641	642	643	644	645	646	647	648	649	650
$\frac{1}{641}$	$\frac{1}{642}$	$\frac{1}{643}$	$\frac{1}{644}$	$\frac{1}{645}$	$\frac{1}{646}$	$\frac{1}{647}$	$\frac{1}{648}$	$\frac{1}{649}$	$\frac{1}{650}$

gals. 3 qts. 1 pt., the greater contains 84 gals. 1 qt. 1 pt.; what is the content of the less? *Ans.* 48 gals. 2 qts.

6. What is the difference between 23 gals. 3 qts. 1 pt. +24 gals. 2 qts. 1 pt.+25 gals. 1 qt. 1 pt.+26 gals. 1 pt. +27 gals. 1 qt., and 28 gals. 1 pt.+29 gals. 1 qt.+30 gals. 3 qts.+31 gals. 1 pt.+32 gals. 3 qts. 1 pt.?
Ans. 24 gals. 3 qts. 1 pt.

7. A gentleman bought a pipe of wine, and having bottled ten dozen (each containing 2 gals. 2 qts. 1 pt.), what remains unbottled? *Ans.* 99 gals. 3 qts.

Examples of Dry Measure.

	bu.	pe.	qts.	pts.		bu.	pe.	qts.	pts.
1.	24	3	5	0	2.	33	3	3	1
	17	3	7	1		25	3	7	1

3. What remains when 244 bu. 3 pe. 7 qts. is taken from 540 bu. 3 pe. 5 qts.? *Ans.* 295 bu. 3 pe. 6 qts.

4. What is the difference between 354 bu. 1 pe. 1 pt., and 256 bu. 6 qts. 1 pt.? *Ans* 98 bu. 2 qts.

5. The crop of a farmer amounted to 1000 bushels, of which 85 bu. 3 pe. 5 qts. 1 pt. became seed, and 250 bu. 2 pe. 4 qts. were used by his family and cattle; what quantity has he now to dispose of?
Ans. 663 bu. 1 pe. 6 qts. 1 pt.

Examples of Time.

	yrs.	mos.	ws.	ds.		yrs.	mos.	ds.
1.	62	4	2	2	2.	24	10	28
	26	4	3	6		18	11	26

3 What is the difference between 24 ds. 23 hrs. 23 m. 29 sec. and 12 ds. 22 hrs. 44 m. 54 sec?
Ans. 12 ds. 38 m. 35 sec.

4. A bond was given for money at interest on the 12th day of the fifth month, 1816, and taken up the 27th of the 1st month, 1818; how long has interest been accumulating? *Ans.* 1 yr. 8 m. 15 ds.

5. From the birth of Washington, 22d Feb., 1732, to

651	652	653	654	655	656	657	658	659	660
$\frac{1}{651}$	$\frac{1}{652}$	$\frac{1}{653}$	$\frac{1}{654}$	$\frac{1}{655}$	$\frac{1}{656}$	$\frac{1}{657}$	$\frac{1}{658}$	$\frac{1}{659}$	$\frac{1}{660}$

the declaration of Independence, 4th July, 1776; and also to the invasion of Baltimore, 12th Sept., 1814, what times respectively had elapsed?

Ans. { To declaration 44 yrs. 4 mos. 11 ds.
To invasion 82 yrs. 6 mos. 19 ds. }

6. John Adams was born June 3d, 1736, and died 4th July, 1826; what was his age at his decease?

Ans. 90 yrs. 1 mo. 1 day.

7. Thomas Jefferson was born April 2d, 1743, and died 4th July, 1826; what was then his age?

Ans. 83 yrs. 3 mos. 2ds.

Note.—When months and days without weeks, are given to be subtracted, let the number of days in the month of the subtrahend be that which you will add to the minuend to make subtraction possible.

Examples of Angular or Circle Measure.

	sig.	°	′	″		sig.	°	′	″
1.	12	00	00	00	2.	6	00	00	00
	8	24	59	58		4	24	56	42

3. When a planet has passed through 7 sig. 20° 24′ 48″ of its orbit, how much is it short of a complete revolution? *Ans.* 4 sig. 9° 35′ 12″.

4. The latitude of Calcutta is 22 deg. 35 min. north, and of Philadelphia 39 deg. 57 min. north; what is their difference of latitude? *Ans.* 17 deg. 22 min.

5. The latitude of Boston is 42 deg. 25 min. north, and of Algiers, 36° 49′ north; what is the difference of their latitudes? *Ans* 5 deg. 36 min.

6. Lizard point, lies 5° $11\frac{3}{4}$′ west of London, and Funchal in Madeira 17° 5′ west; what is their difference of longitude? *Ans.* 11° $53\frac{1}{4}$′.

7. A point on the compass card is 11° 15′, therefore, if 3 points be taken from the quadrant, how many degrees will remain? *Ans.* 56° 15′.

8. The complement of an angle is its defect from 90°, what then is the complement of an angle of 40° 39′?

Ans. 49° 21′.

661	662	663	664	665	666	667	668	669	670
$\frac{1}{661}$	$\frac{1}{662}$	$\frac{1}{663}$	$\frac{1}{664}$	$\frac{1}{665}$	$\frac{1}{666}$	$\frac{1}{667}$	$\frac{1}{668}$	$\frac{1}{669}$	$\frac{1}{670}$

MULTIPLICATION OF MERCANTILE NUMBERS.

This rule briefly shows the product of a quantity composed of several grades of units, as lbs. oz. dwts. grs., applied a given number of times. The reductions of lower grades or denominations to higher are similar to those in Addition of Mercantile Numbers.

If the number of applications, or multiplier, be large, it must be resolved into its factors, or divided into parts, as in the cases below.

Case 1. When the multiplier does not exceed 12.

RULE.—Place it under the lowest grade of the multiplicand; multiply that grade, and reduce the product by division, to units of the grade above it: but since the product may not make exact units of that grade, set down any remainder that may occur, and add the *exact* units to the next product; reduce this product as the first; set down the remainder, if any; carry the quotient to the next grade; and so proceed to the end.

Note 1.—The units' figure of some divisors is a 0, as in shillings, signs, seconds, poles, &c.; in such set down the units' figure of the product as the units of the remainder, dividing only by the tens.

Note 2.—Multiply the numerators of fractions annexed to the lowest grade, and divide the products by their denominators; carry the quotient, and if any remain, write the denominator under it.
The reducing divisors may be seen in addition of M. N.

Examples for the Slate.

Multiply 14 E. 9 dols. 8 d. 6 c. 7 m. by 3; and £5 14s. 8½d. by 12.

	E.	dols.	d.	cts.	m.
1.	14	9	8	6	7
					3
	44	9	6	0	1

	£	s.	d.
2.	5	14	8½
			12
	68	16	6

671	672	673	674	675	676	677	678	679	680
$\frac{1}{671}$	$\frac{1}{672}$	$\frac{1}{673}$	$\frac{1}{674}$	$\frac{1}{675}$	$\frac{1}{676}$	$\frac{1}{677}$	$\frac{1}{678}$	$\frac{1}{679}$	$\frac{1}{680}$

3. Multiply $24 36 cts. by 7, and by 5.
4. Multiply $14 33 cts. 7 m. by 8, and by 6.
5. Multiply 19 dols. 44 cts. by 10, and by 8.
6. Multiply 12 dols. 91 cts. 2 m. by 6, and by 4.
7. Multiply 56 dols. 56 cts. 6 m. by 9, and by 7.
8. Multiply £4 18s. 7½d. by 6, and by 4.
9. Multiply £6 16s. 6d. by 4, and by 8.
10. Multiply 8 dwts. 14 grs. by 6, and by 4.
11. Multiply 3 oz. 10 dwts. 8 grs. by 5, and by 7.
12. Multiply 5 T. 12 cwt. 2 qrs. by 8, and by 4.
13. Multiply 3 qrs. 23 lbs. 13 oz. by 7, and by 3.
14. Multiply 12 lbs. 12 oz. 12 drs. by 9, and by 2.
15. Multiply 7 ʒ 2 ℈ 13 grs. by 10, and by 2.
16. Multiply 2 lbs. 10 ℥ 6 ʒ 1 ℈ by 4, and by 3.
17. Multiply 15 m. 6 fur. 28 po. by 7, and by 5.
18. Multiply 7 fur. 32 po. 4 yds. by 6, and by 3.
19. Multiply 5 yds. 2 ft. 9 in. by 9, and by 3.
20. Multiply 5 E.E. 4 qrs. 3 na. by 4, and by 7.
21. Multiply 24 yds. 3 qrs. 2 na. by 5, and by 7.
22. Multiply 18 E.Fl. 2 qrs. 1 na. by 6, and by 4.
23. Multiply 2 R. 36 po. 28 yds. by 5, and by 4.
24. Multiply 5 A. 3 R. 28 po. by 7, and by 3.
25. Multiply 84 gals. 2 qts. 1 pt. by 8, and by 4.
26. Multiply 4 bu. 3 pe. 5 qts. by 10, and by 2.
27. Multiply 8 mos. 3ws. 4 ds. by 8, and by 4.
28. Multiply 4 ds. 9 hrs. 54 m. by 7, and by 5.
29. Multiply 24° 25′ 50″ by 5

NOTE.—The foregoing questions are answered in the given dividends, Case 1. Division of Mercantile Numbers. There are two multipliers to each of the examples; the single product of their sum will equal the sum of both their products if they are used separately: that is $(mr.+mr.)\times md.=(mr.\times md.)+(mr.\times md.)$; where *mr.* is the contraction of multiplier and *md.* of multiplicand.

Case 2.—When the multiplier exceeds 12, and is a composite number.

RULE.—Resolve the multiplier into its factors; then multiply the quantity by one of the factors, and that product by another, and so on; the last product is the multiple required.

681	682	683	684	685	686	687	688	689	690
$\frac{1}{681}$	$\frac{1}{682}$	$\frac{1}{683}$	$\frac{1}{684}$	$\frac{1}{685}$	$\frac{1}{686}$	$\frac{1}{687}$	$\frac{1}{688}$	$\frac{1}{689}$	$\frac{1}{690}$

The class will first look along the multipliers in the questions below; and resolve each of them into its factors, in answer to the following questions, viz.:

What are the factors of 16? of 36? of 49? of 56? of 64? of 72? of 108? of 20? of 24? of 45? of 63? of 70? of 84? of 96? of 120? of 144? of 49? of 60? of 44? of 121? of 54? of 110? of 28?

Examples for the Slate.

1. 24 tons of hay at 18 dols. 75 cts. per ton?

	dols. cts.		dols. cts.		dols. cts.
	18.75		18.75		18.75
24=	2×12	=	4×6	=	3×8
	37.50		75.00		56.25
	12		6		8
	\$450.00		\$450.00		\$450.00

2. 36 pairs of boots at $4\frac{1}{4}$ dols. per pair? *Ans.* 153 dols.
3. 49 lbs. of cheese, at $9\frac{1}{2}$ cts. per lb? *Ans.* \$4.65$\frac{1}{2}$.
4. 56 pairs silk hose, at \$1.125? *Ans.* 63 dols.
5. 64 pairs shoes at \$1.75? *Ans.* \$112.
6. 72 beaver hats, at \$3.44? *Ans.* \$247.68.
7. 108 barrels of flour, at \$7.20 cts? *Ans.* \$777.6.
8. 20 cwt. of hops, at \$33.6 per cwt.? *Ans.* \$672.
9. 16 cwt. of coffee, at \$10.08 per cwt? *Ans.* \$161.28.
10. 45 ells of calico at 1s. 6d. per ell? *Ans.* £3 7s. 6d.
11. 63 gallons of oil, at \$1.12$\frac{1}{2}$? *Ans.* \$70.87$\frac{1}{2}$.
12. 70 barrels of apples, at \$1 75? *Ans.* \$122.5.
13. 84 bushels of oats, at 35 cents? *Ans.* \$29.40.
14. 96 bushels of corn at $68\frac{3}{4}$ cents? *Ans.* \$66.
15. 120 days' wages, at $87\frac{1}{2}$ cents? *Ans.* \$105.
16. 144 reams of paper at \$3.75 per ream? *Ans.* \$540.
17. Multiply 18 lbs. 7 oz. 11 dwts. 12 grs. by 36.
    Ans. 670 lbs. 8 oz. 14 dwts.
18. Multiply 4 lbs. 3 oz. 10 dwt. 14 grs. by 45.
    Ans. 193 lbs. 2 oz. 16 dwts. 6 grs.
19. Multiply 4 T. 12 cwt. 2 qrs. 16 lbs. by 49.
    Ans. 226 T. 19 cwt. 2 qrs.

691	692	693	694	695	696	697	698	699	700
$\frac{1}{691}$	$\frac{1}{692}$	$\frac{1}{693}$	$\frac{1}{694}$	$\frac{1}{695}$	$\frac{1}{696}$	$\frac{1}{697}$	$\frac{1}{698}$	$\frac{1}{699}$	$\frac{1}{700}$

20. Multiply 3 cwt. 3 qrs. 23 lbs. 13 oz. by 60.
Ans. 237 cwt. 3 qrs. 12 oz.

21. Multiply 5 lbs. 4 ℥ 7 ʒ 2 ℈ 18 grs. by 64.
Ans. 346 lbs. 7 oz. 5 drs. 2 sc. 12 grs.

22. Multiply 1 lb. 1 ℥ 1 ʒ 1 ℈ 1 gr. by 72.
Ans. 79 lbs. 1 dr. 12 grs.

23. Multiply 22 m. 7 fur. 32 po. 4 yds. by 56.
Ans. 1286 m. 5 fur. 32 po. 4 yds.

24. Multiply 5 yds. 2 ft. 9⅔ in. by 63.
Ans. 373 yds. 2 ft. 9 in.

25. Multiply 2 E.E. 2 qrs. 1 na. by 108.
Ans. 264 E.E. 3 qrs.

26. Multiply 4 yds. 3 qrs. 2 na. by 44.
Ans. 214 yds. 2 qrs.

27. Multiply 5 E.Fl. 2 qrs. 3 na. by 121.
Ans. 715 E.Fl. 2 qrs. 3 na.

28. Multiply 3 A. 3 R. 33 po. 23 yds. by 54.
Ans. 213 A. 3 R. 23 po. 1¾ yds.

29. Multiply 2 A. 2 R. 22 po. 25 yds. by 108.
Ans. 285 A. 1 R. 25 po. 7¾ yds.

30. Multiply 2 T. 2 hhds. 42 gals. by 64.
Ans. 170 T. 2 hhds. 42 gals.

31. Multiply 3 hhds. 33 gals. 3 qts. by 110.
Ans. 388 hhds. 58 gals. 2 qts.

32. Multiply 4 bu. 3 pe. 7 qts. by 96. *Ans.* 477 bu.

33. Multiply 8 bu. 2 pe. 5 qts. 1 pt. by 54.
Ans. 468 bu. 1 pe. 1 qt.

34. Multiply 5 years 4 mos. 3 ws. 5 ds. by 49.
Ans. 263 yrs. 7 mo. 2 w.

35. Multiply 4 ds. 9 hrs. 54 m. 36 sec. by 24.
Ans. 105 ds. 21 hrs. 50 m. 24 sec.

36. Multiply 12 deg. 50 m. 25 sec. by 28.
Ans. 359° 31′ 40″.

Case 3. When the multiplier is not a rectangle of any two factors in the table of multiples of the first twelve numbers, and is not too large.

RULE. Use the two factors which produce the nearest

701	702	703	704	705	706	707	708	709	710
$\frac{1}{701}$	$\frac{1}{702}$	$\frac{1}{703}$	$\frac{1}{704}$	$\frac{1}{705}$	$\frac{1}{706}$	$\frac{1}{707}$	$\frac{1}{708}$	$\frac{1}{709}$	$\frac{1}{710}$

rectangle to it, be it less or greater; then add, if less, or subtract if greater, the product of the difference.

The class will now look along the line of multipliers below, and answer the following questions:

What is the nearest rectangle to 33? and what is the difference between it and 33? The same in regard to 53? to 29? to 34? to 41? to 19? to 39? to 94? to 26? to 47? to 65?

Examples for the Slate.

1. 33 yards of cassimer at $1.75 per yard?
2. 29 pairs of boots at £1 2s. 9d. per pair?

```
      dols. cts.                         £  s. d.
       1.75×1                            1  2  9×1.
 33=     4×8+1                    29=       5×6—1
       ─────                          ─────────
       7.00 price of 4 yds.           5 13  9 price of 5 prs.
          8                                 6
      ──────                          ─────────
      56.00 price of 8×4.            34  2  6 price of 6×5.
    +  1.75 price of 1.            —  1  2  9 price of 1.
      ──────                          ─────────
     $57.75 price of 33.            £32 19  9 price of 29.
     ══════                          ═════════
```

3. Multiply 4 lbs. 8 oz. 12 dwts. by 53.
   Ans. 249 lbs. 11 oz. 16 grs.
4. Multiply 1 cwt. 2 qrs. 18 lbs. by 34.
   Ans. 56 cwt. 1 qr. 24 lbs.
5. Multiply 2 qrs. 10 lbs. by 41.
   Ans. 24 cwt. 18 lbs.
6. Multiply 5 lbs. 10 oz. 6 drs. 2 sc. by 19.
   Ans. 112 lbs. 1 oz. 6 drs. 2 sc.
7. Multiply 29 m. 5 fur. 28 po. by 39.
   Ans. 1158 m. 6 fur. 12 po.
8. Multiply 18 yds. 3 qrs. 2 na. by 94.
   Ans. 1774 yds. 1 qr.
9. Multiply 1256 A. 3 R. 20 po. by 26.
   Ans. 32678 A. 3 R.
10. Multiply 29 gals. 2 qts. by 47.
    Ans. 1386 gals. 2 qts.

711	712	713	714	715	716	717	718	719	720
$\frac{1}{711}$	$\frac{1}{712}$	$\frac{1}{713}$	$\frac{1}{714}$	$\frac{1}{715}$	$\frac{1}{716}$	$\frac{1}{717}$	$\frac{1}{718}$	$\frac{1}{719}$	$\frac{1}{720}$

11. Multiply 3 bu. 2 pe. 2 qts. by 65.
Ans. 231 bu. 2 pe. 2 qts.

Case 4.—When the multiplier is any high number.

RULE.—First use in succession, the decimal factors of the highest rank; namely, 10 and 10 for 100, and 10 again for 1000, &c., also the figure of that rank, if more than a unit; to the result add the product of the several lower ranks of the multiplier into the corresponding places of the operation, as in the first example below.

1. What is the value of 365 lbs. of butter at 1s. 7½d. per lb.?

£	s.	d.		
0	1	7½×5		
		10×10×3=300,		300+60+5=365
0	16	3×6		10×10×3=300
		10		10×6= 60
				1×5= 5
8	2	6		
		3		
24	7	6	product of 300.	
4	17	6	" of 60.	
	8	1½	" of 5.	
Ans. 29	13	1½	" of 365.	

2. Multiply £3 8s. 11¼d. by 739.
Ans. £2547 4s. 9¾d.

3. Multiply 4 lbs. 8 oz. 12 dwts. 18 grs. by 636.
Ans. 3001 lbs. 9 oz. 9 dwts.

4. Multiply 1 cwt. 2 qrs. 18 lbs. by 408.
Ans. 677 cwt. 2 qrs. 8 lbs.

5. Multiply 2 qrs. 10 lbs. by 492.
Ans. 289 cwt. 3 qrs. 20 lbs.

6. Multiply 5 lbs. 10 oz. 6 drs. 18 grs. by 228.
Ans. 1346 lbs. 6 ℥ 4 ʒ 1 ℈ 4 grs.

7. Multiply 29 m. 5 fur. 28 po. by 468.
Ans. 13905 m. 3 fur. 24 po.

721	722	723	724	725	726	727	728	729	730
$\frac{1}{721}$	$\frac{1}{722}$	$\frac{1}{723}$	$\frac{1}{724}$	$\frac{1}{725}$	$\frac{1}{726}$	$\frac{1}{727}$	$\frac{1}{728}$	$\frac{1}{729}$	$\frac{1}{730}$

8. Multiply 18 yds. 3 qrs. 2 na. by 1128.
Ans. 21291 yards.
9. Multiply 62 A. 3 R. 15 po. by 1040.
Ans. 65357 A. 2 R.
10. Multiply 29 gals. 2 qts. by 564.
Ans. 16638 gals.
11. Multiply 3 bu. 2 pe. 2 qts. by 780.
Ans. 2778 bu. 3 pe.

DIVISION OF MERCANTILE NUMBERS.

This rule teaches to divide a sum, or quantity composed of units of different grades into any number of equal parts. The reductions of higher grades of units to lower, by multiplication of remainders, are shown in the following

Class Exercise on the Rules of Reduction.

How are dollars reduced to dimes? dimes to cents? cents to mills? and so on through all the denominations.

U. S. Money.

$×10=dimes.
di.×10=cents.
cts.×10=mills.
$×100=cts.

Currencies.

£×20=s.
s.×12=d.

Troy Weight.

lbs.×12=oz.
oz.×20=dwts.
dwts.×24=grs.

Avoir. Weight.

T×20=cwt.
cwt.×4=qrs.
qrs.×28=lbs.
lbs.×16=oz.
oz.×16=drs.

Apoth. Weight.

lbs.×12=oz.
oz.×8=drs.
dr.×3=sc.
sc.×20=grs.

Long Measure.

deg.×69½=m.
m.×8=fur.
fur.×40=po.
po.×5½=yds.
yds.×3=ft.
ft.×12=in.

Cloth Measure.

E.E.×5=qrs.
yds.×4=qrs.
E.Fl.×3=qrs.
E.H.×2½=qrs.
qrs.×4=na.
na.×2¼=in.

Square Measure.

Acres×4=R.
Roods×40=sq. po.
sq. po.×30¼=sq. yds.
sq. yds.×9=sq. ft.
sq. ft.×144=sq. in.

Liquid Measure.

T.×252=gals.
Pi.×126=gals.

731	732	733	734	735	736	737	738	739	740
$\frac{1}{731}$	$\frac{1}{732}$	$\frac{1}{733}$	$\frac{1}{734}$	$\frac{1}{735}$	$\frac{1}{736}$	$\frac{1}{737}$	$\frac{1}{738}$	$\frac{1}{739}$	$\frac{1}{740}$

pun.×84=gals.
hhds.×63=gals.
tier.×42=gals.
bls.×31½=gals.
gals.×4=qts.
qts.×2=pts.
pts.×4=gls.

Dry Measure.

bu.×4=pe.
pe.×8=qts.
qts.×2=pts.

Time.

yrs.×12=Cal. mo.
yrs.×13=L. mo.
mos.×30=ds.
ws.×7=ds.
ds.×24=hrs.
hrs.×60=m.
m.×60=sec.

Angular or Circle Measure.

Circ.×12=signs.
signs×30=deg.
deg.×60=m.
m.×60=sec.

Cubic Measure.

Cu. yd.×27=cu. ft.
Cu. ft.×1728=cu. in.

Every operation of this kind may take the form of Long Division; exhibiting on the slate the reduction of remainders to units of lower grades, with the several parts of the work. But when the divisor is small, and since the remainder should be smaller, the reduction may be performed mentally, and the quotient written under the dividend.

Case. 1. When the divisor does not exceed 12.

RULE. Divide the highest grade of units in the given quantity, as in Simple Division; if a remainder occur, reduce it, by the proper multiplier, to units of the grade below, adding the *given* units of that grade to the product: divide this result as at the first; reduce the remainder; add, and continue the same process to the end.

Place the quotient under the dividend, and the divisor under the last remainder.

Note 1. When the reducing multiplier is 10, 20, 30, 40, or 60, as it may sometimes be, multiply by the tens' figure only, and divide, add the tens of the next grade, prefix the remainder, and proceed.

Note 2. When the quantity has a fraction at the last, include it as a part of the remainder, subscribing the divisor.

Examples for the Slate.

Divide 449 dols. 60 c. 1 m. by 3, and £68 16s. 6d. by 12.

	dols. cts. m.		£ s. d.
1.	3)449 60 1	2.	12)68 16 6
	149 86 7		5 14 8½

741	742	743	744	745	746	747	748	749	750
$\frac{1}{741}$	$\frac{1}{742}$	$\frac{1}{743}$	$\frac{1}{744}$	$\frac{1}{745}$	$\frac{1}{746}$	$\frac{1}{747}$	$\frac{1}{748}$	$\frac{1}{749}$	$\frac{1}{750}$

The questions in Case 1, Multiplication of Mercantile Numbers, are answers to the following collection.

3. Divide 292.32 dols. by 12.
4. Divide 200.718 dols. by 14.
5. Divide 349.92 dols. by 18.
6. Divide 129.12 dols. by 10.
7. Divide 905.056 dols. by 16.
8. Divide £49 6s. 3d. by 10.
9. Divide £81 18s. by 12.
10. Divide 4 oz. 5 dwts. 20 grs. by 10.
11. Divide 3 lbs. 6 oz. 4 dwts. by 12.
12. Divide 67 T. 10 cwt. by 12.
13. Divide 9 cwt. 2 qrs. 14 lbs. 2 oz. by 10.
14. Divide 5 qrs. 12 oz. 4 drs. by 11.
15. Divide 11 ℥ 6 ʒ 1 ℈ 16 gr. by 12.
16. Divide 20 lbs. 3 ℥ 4 ʒ 1 ℈ by 7.
17. Divide 190 m. 16 po. by 12.
18. Divide 70 fur. 14 po. 3 yds. by 9.
19. Divide 71 yards by 12.
20. Divide 65 E.E. 2 qrs. 1 na. by 11.
21. Divide 298 yds. 2 qrs. by 12.
22. Divide 187 E.Fl. 1 qr. 2 na. by 10.
23. Divide 26 R. 12 po. 10 yds. by 9.
24. Divide 59 A. 1 R. by 10.
25. Divide 1015 gals. 2 qts. by 12.
26. Divide 58 bu. 3 pe. 4 qts. by 12.
27. Divide 106 mos. 2 ws. 6 ds. by 12.
28. Divide 52 ds. 22 hrs. 48 m. by 12.
29. Divide 122° 9′ 10″ by 5.

Case 2. When the divisor is a composite number and its factors small.

RULE. Divide the given quantity by one of the factors, as in Case 1, and divide that quotient by the other factor: the last quotient is the thing required.

The class will now look along the line of divisors in the examples below, and prepare to give oral responses to the following questions; viz.:

What are the factors of 21? of 24? of 36? of 48? of

751	752	753	754	755	756	757	758	759	760
$\frac{1}{751}$	$\frac{1}{752}$	$\frac{1}{753}$	$\frac{1}{754}$	$\frac{1}{755}$	$\frac{1}{756}$	$\frac{1}{757}$	$\frac{1}{758}$	$\frac{1}{759}$	$\frac{1}{760}$

63? of 88? of 20? of 56? of 96? of 100? of 27? of 64? of 54? of 99? of 81? of 120? of 42? of 108? of 132? of 144? of 121? of 84? of 110? of 60?

Examples for the Slate.

1. If 16 cwt. of cheese cost $128.64, how much was that for 1 cwt.? how much per lb.?

```
           dols. cts.                    dols. cts.
16=4×4)128.64              112=28×4)8.04
        ------                      ----
      4)32.16                28=7×4)2.0100
        -----                       ------
Ans.  $8.04 per cwt.               7) .5025
                                      -----
                         Ans.  $.0717 6/7 per lb.
```

2. Divide $254.964 by 21.
   Ans. 12 dols. 14 cts. $1\frac{1}{7}$ m.
3. Divide $546.986 by 24.
   Ans. 22 dols. 79 cts. $1\frac{1}{12}$ m.
4. Divide $845.72 by 36.
   Ans. 23 dols. 4 d. $9\frac{2}{9}$ cts.
5. Divide $792.288 by 48.
   Ans. 16 dols. 50 cts 6 m.
6. Divide $118.548 by 63.
   Ans. 1 dol. 88 cts. $1\frac{5}{7}$ m.
7. Divide $256.35 by 88.
   Ans. 2 dols. 91 cts. $3\frac{3}{44}$ m.
8. Divide £150 6s. 8d. by 20. *Ans.* £7 10s. 4d.
9. Divide £98 8s. by 36. *Ans.* £2 14s. 8d.
10. Divide £71 13s. 10d. by 56. *Ans.* £1 5s. $7\frac{1}{4}$d.
11. Divide £44 4s. by 96. *Ans.* 9s. $2\frac{1}{2}$d.
12. Divide £31 10s. by 100. *Ans.* 6s. $3\frac{3}{5}$d.
13. Divide 14 lbs. 10 oz. 13 dwts. by 27.
    Ans. 6 oz. 12 dwts. 8 grs.
14. Divide 48 lbs. 6 oz. 6 dwts. by 48.
    Ans. 1 lb. 2 dwts. 15 grs.
15. Divide 19 T. 19 cwt. 3 qrs. 12 lbs. by 36.
    Ans. 11 cwt. 12 lbs.

761	762	763	764	765	766	767	768	769	770
$\frac{1}{761}$	$\frac{1}{762}$	$\frac{1}{763}$	$\frac{1}{764}$	$\frac{1}{765}$	$\frac{1}{766}$	$\frac{1}{767}$	$\frac{1}{768}$	$\frac{1}{769}$	$\frac{1}{770}$

16. Divide 25 cwt. 2 qrs. 24 lbs. by 64.
Ans. 1 qr. 17 lbs.
17. Divide 42 lbs. 10 ℥ 6 ʒ 1 ℈ 4 grs. by 54.
Ans. 9 ℥ 4 ʒ 16 grs.
18. Divide 102 lbs. 10 ℥ by 99.
Ans. 1 lb. 3 ʒ 2 ℈ $3\frac{1}{33}$ grs.
19. Divide 45 m. 4 fur. 20 po. by 81.
Ans 4. fur. 20 po.
20. Divide 249 m. 6 fur. by 120. *Ans.* 2 m. 26 po.
21. Divide 149 E.E. 2 na. by 42.
Ans. 3 E.E. 2 qrs. 3 na.
22. Divide 560 yds. 1 qr. by 108. *Ans.* 5 yds. 3 na.
23. Divide 473 E.Fl. by 132.
Ans. 3 E.Fl. 1 qr. 3 na.
24. Divide 598 A. 2 R. by 144. *Ans.* 4 A. 25 po.
25. Divide 547 A. 3 R. 4 po. by 121.
Ans. 4 A. 2 R. 4 po. 10 yds.
26. Divide 456 hhds. 48 gals. by 96.
Ans. 4 hhds. 47 gals. 3 qts.
27. Divide 954 hhds. 21 gals. by 84.
Ans. 11 hhds. 22 gals. 3 qts.
28. Divide 1001 bu. 6 qts. by 108.
Ans. 9 bu. 1 pe. $1\frac{8}{27}$ pts.
29. Divide 500 bu. 2 pe. by 110. *Ans.* 4 bu. $2\frac{1}{5}$ pe.
30. Divide 1817 years 5 mos. by 60.
Ans. 30 yrs. 3 mos. 14 ds. 12 hrs.
31. Divide 360° by 24. *Ans.* 15°.

This last example shows how to reduce circle measure to time: for as the day is 24 hours, and the circle 360°, the answer shows that 15°=1 hour of time.

Case 3. When the divisor is any number whatever.

Rule. In this case, the quotient is placed on the right of the dividend, leaving the space below for the exhibition of the work, as in Long Division. Divide the highest grade of units first; reduce the remainder, if any, to units of the grade below, adding the units of that grade which may be in the given quantity to the product; divide the result; and continue the same process to the end.

771	772	773	774	775	776	777	778	779	780
$\frac{1}{771}$	$\frac{1}{772}$	$\frac{1}{773}$	$\frac{1}{774}$	$\frac{1}{775}$	$\frac{1}{776}$	$\frac{1}{777}$	$\frac{1}{778}$	$\frac{1}{779}$	$\frac{1}{780}$

Examples for the Slate.

1. Divide 178 cwt. 3 qrs. 14 lbs. by 53.

```
       cwt.  qrs.  lbs.
    53)178   3    14(3 cwt. 1 qr. 14 bs.
       159
       ---
Reduce rem. to qrs. 19 cwt. and add 3 qrs.
        4
    53)79(1 qr.
       53
       --
Reduce rem. to lbs. 26 qrs. and add 14 lbs.
       28
      ---
      212
       52
      ---
   53)742(14 lbs.
       53
      ---
      212
      212
      ===
```

2. Divide £39 14s. 5¼d. by 57. *Ans.* 13s. 11¼d.
3. Divide £125 4s. 9d. by 43. *Ans.* £2 18s. 3d.
4. Divide 3001 lbs. 9 oz. 9 dwts. by 636.
Ans. 4 lbs. 8 oz. 12 dwts. 18 grs.
5. Divide 289 cwt. 3 qrs. 20 lbs. by 492.
Ans. 2 qrs. 10 lbs.
6. Divide 144 m. 4 fur. 20 po. 5 ft. by 39.
Ans. 3 m. 5 fur. 26 po. 2 ft. 8 in.
7. Divide 13905 m. 3 fur. 24 po. by 468.
Ans. 29 m. 5 fur. 28 po.
8. Divide 534 yds. 2 qrs. 2 na. by 47.
Ans. 11 yds. 1 qr. 2 na.
9. Divide 21291 yds. by 1128. *A.* 18 yds. 3 qrs. 2 na.
10. Divide 77 A. 1 R. 33 po. by 51.
Ans. 1 A. 2 R. 3 po.

781	782	783	784	785	786	787	788	789	790
$\frac{1}{781}$	$\frac{1}{782}$	$\frac{1}{783}$	$\frac{1}{784}$	$\frac{1}{785}$	$\frac{1}{786}$	$\frac{1}{787}$	$\frac{1}{788}$	$\frac{1}{789}$	$\frac{1}{790}$

11. Divide 65357 A. 2 R. by 1040.
Ans. 62 A. 3 R. 15 po.
12. Divide 1811 gals. 7 pts. by 65.
Ans. 27 gals. 7 pts.
13. Divide 16638 gals. by 584. *Ans.* 29 gals. 2 qts.
14. Divide 467 bu. 1 pe. 6 qts. by 54.
Ans. 8 bu. 2 pe. 5 qts.
15. Divide 2778 bu. 3 pe. by 780.
Ans. 3 bu. 2 pe. 2 qts.
16. Divide 206 months 4 days by 26.
Ans. 7 mos. 3 ws. 5 ds.

REDUCTION OF MERCANTILE NUMBERS.

This operation proceeds by multiplication and division to change the grade of the units in a quantity without affecting the value of the quantity.

Rules.

1. When the reduction is from a higher to a lower grade: Let the number of the units be increased by multiplication, and if there be lower units in the given quantity, add them to the product.

2. When the reduction is from a lower to a higher grade: Let the number of the units be diminished by division, and let the units which remain, if any, being a part of the dividend, retain their former value.

The numbers used as multipliers and divisors in these rules, are the mercantile numbers found in the several tables of measures, weights, &c.

Note. The answers to several of the following questions are given in the questions annexed: the questions mutually answer each other.

791	792	793	794	795	796	797	798	799	800
$\frac{1}{791}$	$\frac{1}{792}$	$\frac{1}{793}$	$\frac{1}{794}$	$\frac{1}{795}$	$\frac{1}{796}$	$\frac{1}{797}$	$\frac{1}{798}$	$\frac{1}{799}$	$\frac{1}{800}$

Examples in Currencies and Sterling Money.

1. Change £234 15s. 7d. to pence, and again to £.

```
      £   s.  d.                     pence.
     234  15  7               1s=12)56347
 1£=  20                           ------
    -----                      2,0)469,5s. 7d.
    4695s.  including 15s.          ----
 1s.=  12                     Ans. 234£ 15s. 7d.
    -----                     ================
Ans. 56347d. including 7d.
    ======
```

2. Reduce £24 and £40 19s. each to shillings.
3. Reduce 480s. and 819s. each to pounds.
4. Reduce £51 13s. and £55 9s. each to shillings.
5. Reduce 1033s. and 1109s. each to pounds.
6. Reduce 14s. and 7s. 9d. each to pence.
7. Reduce 168d. and 93d. each to shillings.
8. Reduce £80 and £97 each to parcels of three pence.
9. Reduce 6400 and 7760 parcels of three pence each to pounds.

Examples in Troy Weight.

1. Reduce 10 dwts. 18 grs. and 5 dwts. 6 grs. each to grains.
2. Reduce 258 grains and 126 grains each to dwts.
3. Reduce 2 oz. 4 dwts. and 9 lbs. each to dwts.
4. Reduce 44 dwts. and 2160 dwts. each to oz.
5. The eagle weighs 10 dwts. 18 grs. of standard gold, the half eagle 5 dwts. 9 grs. and the quarter eagle 2 dwts. $16\frac{1}{2}$ grs.; how many grains will 100 of each weigh?
6. How many eagles, half eagles, and quarter eagles, and of each an equal number, will 45150 grains of standard gold make?

Examples in Avoirdupois Weight.

1. In 17 T. 18 cwt. 2 qrs. 25 lbs. 11 oz. 14 drs., how many drachms? *Ans.* 10285502 drs.
2. In 16800 lbs. how many tons? *Ans.* 7 T. 10 cwt.
3. In 180 boxes of raisins, each 15 lbs. how many cwt.?

801	802	803	804	805	806	807	808	809	810
$\frac{1}{801}$	$\frac{1}{802}$	$\frac{1}{803}$	$\frac{1}{804}$	$\frac{1}{805}$	$\frac{1}{806}$	$\frac{1}{807}$	$\frac{1}{808}$	$\frac{1}{809}$	$\frac{1}{810}$

4. How many parcels of 15 lbs. are in 24 cwt. 12 lbs.?
5. Reduce 8 cwt. 2 qrs. 17 lbs. to pounds.
6. Reduce 969 lbs. to cwts.

Examples in Apothecaries' Weight.

1. Reduce 12 lbs. to oz. drs. ℈ grs.
Ans. 144 ℥ 1152 ʒ 3456 ℈ 69120 grs.
2. In 69120 grains, how many pounds?
3. In 6 oz. 5 drs. 2 ℈ 12 grs., how many parcels of 8 grains each? *Ans.* 404 parcels.
4. In 202 parcels of 16 grains each, how many ounces, drachms, scruples and grains?
Ans. 6 oz. 5 drs. 2 ℈ 12 grs.

Examples in Long Measure.

1. Reduce 10 m. 3 fur. 8 po. to inches.
Ans. 658944.
2. In 68167 inches, how many miles, &c.
Ans. 1 m. 24 po. 1 yd. 1 ft. 7 in.
3. The distance from Washington to Baltimore is 38 miles; how many revolutions of a wheel 3 yards in circumference will it require to measure it? *Ans.* 22293$\frac{1}{3}$.
4. What distance will a wheel of 16 ft. 6 in. measure in turning 640 times? *Ans.* 2 miles.
5. What is the Earth's circumference in yards?
Ans. 44035200.

Examples in Cloth Measure.

1. In 25 yards 2 qrs. 3 na., how many nails?
Ans. 411.
2. In 4560 nails, how many yards? *Ans.* 285.
3. Reduce 33 E.Fl. 2 qrs. 2 na. to nails. *Ans.* 406.
4. In 5648 nails, how, many E.Fl.?
Ans. 470 E.Fl. 2 qrs.
5. In ten pieces of cloth, 24 yds. 2 qrs. 2 na. each, how many ells English? *Ans.* 197 E.E.
6. In 35 E.E. 3 qrs. how many yds.?
Ans. 44 yds. 2 qrs.

811	812	813	814	815	816	817	818	819	820
$\frac{1}{811}$	$\frac{1}{812}$	$\frac{1}{813}$	$\frac{1}{814}$	$\frac{1}{815}$	$\frac{1}{816}$	$\frac{1}{817}$	$\frac{1}{818}$	$\frac{1}{819}$	$\frac{1}{820}$

Examples in Square Measure.

1. Reduce 24 A. 2 R. 16 po. 28 yds. to yds. *Ans.* 119092 yds.
2. In 119092 sq. yds. how many acres, roods, &c.? *Ans.* 24 A. 2 R. 16 po. 28 yds.
3. Reduce 18 yds. 8 ft. 84 in. to inches. *Ans.* 24564 in.
4. In 24564 sq. in. how many yards? *Ans.* 18 yds. 8 ft. 84 in.
5. A tract of land containing 560 A. 2 R. 35 po. is to be divided equally among three brothers, how many square yards should each man take? *Ans.* 904626$\frac{1}{4}$ yds.
6. Reduce 904626$\frac{1}{4}$ square yards to acres. *Ans.* 186 A. 3 R. 25 po.

Examples in Liquid Measure.

1. Reduce 27 T. 3 hhds. 53 gals. 5 pts. to pints. *Ans.* 56373 pts.
2. Reduce 28186 quarts to tons. *Ans.* 27 T. 3 hhds. 53 gals. 2 qts.
3. One dozen bottles of wine is usually reckoned to contain 10$\frac{1}{2}$ quarts, then how many such dozens may be filled out of 5 hhds? *Ans.* 120 dozens.
4. How many barrels of 31$\frac{1}{2}$ gallons, and half barrels of 16 gals., and of each an equal number, may be filled out of 3325 gallons of ale? *Ans.* 70 of each.

Examples in Dry Measure.

1. In 45 bu. 2 pe. 6 qts. 1 pt., how many pints? *Ans.* 2925.
2. In 2899 pints, how many bushels, pecks, &c. *Ans.* 45 bu. 1 pe. 1 qt. 1 pt.
3. How many 3 bushel sacks will 8640 pints of wheat fill? *Ans.* 45 sacks.
4. Reduce 5850 pints of wheat to bushels? *Ans.* 91 bu. 1 pe. 5 qts.

821	822	823	824	825	826	827	828	829	830
$\frac{1}{821}$	$\frac{1}{822}$	$\frac{1}{823}$	$\frac{1}{824}$	$\frac{1}{825}$	$\frac{1}{826}$	$\frac{1}{827}$	$\frac{1}{828}$	$\frac{1}{829}$	$\frac{1}{830}$

Examples in Time.

1. How many seconds in a solar year, or 365 days 5 hrs. 48 m. 48 sec.? *Ans.* 31556928.

2. In a lunar month, or 29 days 12 hrs. 44 m. 3 sec., how many oscillations of the pendulum? *Ans.* 2551443.

Examples in Angular or Circle Measure.

1. How many seconds in a sign of the Zodiac? *Ans.* 108000.

2. How many signs are in 648000 seconds? *Ans.* 6 signs.

To Reduce Inferior to Decimal parts of Superior Grades.

In addition and multiplication of mercantile numbers we have reduced units of lower grades to higher by division; we called the same quantities by units of other names; *one* we took for 2, for 3, for 4, 6, 8, 9, 10, 12, 16, 20, 24, 27, 30, 40, 60, or any other number, as the tables direct. But in all these divisions the quantity divided was some multiple of the divisor; and if not, we set it down in its place, under its own title, without any change; or if it exceeded a multiple of the divisor, we set down the excess, or remainder, in like manner. It may have been noticed also, that the number under any inferior title in a quantity, is less than the tabular number of that title.

Now it is desirable to be rid of so many titles to one quantity, and to have all its parts brought under the title of its highest or measuring unit; the unit of which the price is usually given. For this purpose the forms below are presented, which recapitulate the reductions of addition and multiplication.

In class: How are pence reduced to shillings? shil. to £s? drs. to oz.? oz. to lbs.? lbs. to qrs.? qrs. to cwts.? cwts. to T.? And so through all the denominations. See next page.

831	832	833	834	835	836	837	838	839	840
$\frac{1}{831}$	$\frac{1}{832}$	$\frac{1}{833}$	$\frac{1}{834}$	$\frac{1}{835}$	$\frac{1}{836}$	$\frac{1}{827}$	$\frac{1}{838}$	$\frac{1}{839}$	$\frac{1}{840}$

12)d.	4)ws.	60)sec.	4)qrs.	acres.
20)s.	13)mos.	60)m.	yds.	—
£	yrs.	30)deg.	—	
—	—	12)signs.		4)gls.
16)drs.	24)grs.	cir.	2)pts.	2)pts.
16)oz.	20)dwts.	—	8)qts.	4)qts.
28)lbs.	12)oz.	12)in.	4)pe.	gals.
4)qrs.	lbs.	3)ft.	bu.	—
20)cwts.	—	5½)yds.	—	
T.	20)grs.	40)po.		30)ds.
—	3)℈	8)fur.	144)sq. in.	c. m.
60)sec.	8)ʒ	m.	9)sq. ft.	—
60)m.	12)℥	—	30¼)sq. yds.	
24)hrs.	lbs.	2¼)in.	40)sq. po.	365¼)ds.
7)ds.	—	4)na.	4)roods.	yrs.

And since, in Division of Decimals, page 44, we have seen the manner of increasing the parts of a less number so as to make possible the division of it by a greater, all the inferior titles of a quantity may be reduced to the superior by the following

Rule. Set the given names one under another, in the order of their grades; beginning at the inferior, divide each by its proper tabular number, annexing ciphers, and setting the quotient on the right of the number next below, and prefixing the units' point.

Examples for the Slate.

1. Reduce 17s. 4½d. to the decimal of a £.

½) 1.0
12) 4.500d.
20)17.3750s.
Ans. 0.86875 £.

2. Reduce 12s. 6d. to the decimal of a £.

12) 6.0d.
20)12.50s.
Ans. 0.625£.

3. Reduce 4½d. to the decimal of a shilling. *Ans.* .375s.

4. Reduce 9d. to the decimal of a £. *Ans.* .0375£.

5. Reduce 5 oz. 12 dwts. 15.744 grs to the decimal of a pound troy. *Ans.* .4694 lb.

841	842	843	844	845	846	847	848	849	850
$\frac{1}{841}$	$\frac{1}{842}$	$\frac{1}{843}$	$\frac{1}{844}$	$\frac{1}{845}$	$\frac{1}{846}$	$\frac{1}{847}$	$\frac{1}{848}$	$\frac{1}{849}$	$\frac{1}{850}$

6. Reduce 8 oz. 8 dwts. to decimal parts of a lb. troy. *Ans.* .7 lb.
7. Reduce 10 oz. 4 dwts. to decimal parts of a lb. troy. *Ans.* .85 lb.
8. Reduce 4 cwt. 2 qrs. to decimal parts of a ton. *Ans.* .225 T.
9. Reduce 2 qrs. 14 lbs. to decimal parts of a cwt. *Ans.* .625 cwt.
10. Reduce 7 drachms to decimal parts of a lb. apoth. *Ans.* .0729 lb.
11. Reduce 7.3 grains to decimal parts of a scruple. *Ans.* .365 sc.
12. Reduce 10 oz. to decimal parts of a lb. apoth. *Ans.* .83⅓ lb.
13. Reduce 2 na. 1.5 in. to decimal parts of a yard. *Ans.* .16666+yd.
14. Reduce 6 fur. 8.64 po. to decimal parts of a league. *Ans.* .259 L.
15. Reduce 6 fur. 4 po. to decimal parts of a mile. *Ans.* .7625 m.
16. Reduce 3 qrs. 2 na. to decimal parts of a yard. *Ans.* .875 yd.
17. Reduce 4 qrs. 2 na. to decimal parts of an E.E. *Ans.* .9 E.E.
18. Reduce 1 R. 14 po. to decimal parts of an acre. *Ans.* .3375 A.
19. Reduce 26.8 po. to decimal parts of a rood. *Ans.* .67 rood.
20. Reduce 15 gals. 3 qts. to the decimal of a hhd. *Ans.* .25 hhd.
21. Reduce 3 qts. 1 pt. to decimal parts of a gallon. *Ans.* .875 gal.
22. Reduce 6 qts. 1 pt. to decimal parts of a peck. *Ans.* .8125 pe.
23. Reduce 1 pe. 4 qts. to decimal parts of a bushel. *Ans.* .375 bu.
24. Reduce 15 hrs. to decimal parts of a day. *A.* .625 d.
25. Reduce 146 ds. 2 hrs. 24 minutes to decimal parts of a year of 365¼ days. *Ans.* .4 yr.

851	852	853	854	855	856	857	858	859	860
$\frac{1}{851}$	$\frac{1}{852}$	$\frac{1}{853}$	$\frac{1}{854}$	$\frac{1}{855}$	$\frac{1}{856}$	$\frac{1}{857}$	$\frac{1}{858}$	$\frac{1}{859}$	$\frac{1}{860}$

To Reduce Decimal Parts of Superior to Units of Inferior Grade.

RULE. Multiply the given parts by the proper tabular nnmber, as if they were multiples of the measuring unit of the number, and point off in the product as many ranks of parts as there are in both factors, as taught in Multiplication of Decimals.

Examples for the Slate.

1. Reduce .625£ to units of inferior grade; also, .375s.

```
    .625£              .375s.
 1£= 20s.          1s.= 12d.
   ______             ______
  12.5|00s.          4.5|00
  12                 ======
   ___              Ans. 4.5=4 5/10=4 1/2d.
   6.0d.  Ans. 12s. 6d.
  =====
```

2. Reduce .75£ to units of shillings. *Ans.* 15s.
3. Reduce .495 lb. troy to units of oz. dwts. grs. *Ans.* 5 oz. 18 dwts. 19.2 grs.
4. Reduce .7 lb. troy to units of oz. and dwts. *Ans.* 8 oz. 8 dwts.
5. Reduce .625 cwt. to units of qrs. and lbs. *Ans.* 2 qrs. 14 lbs.
6. Reduce .225 ton to units of cwts. and qrs. *Ans.* 4 cwt. 2 qrs.
7. Reduce .0375£ to units of pence. *Ans.* 9d.
8. Reduce .02734375 lb. avoir. to units of drachms. *Ans.* 7 drs.
9. Reduce .85 lb. troy to units of oz. and dwts. *Ans.* 10 oz. 4 dwts.
10. Reduce .365 ℈ to units and tenths of a grain. *Ans.* 7.3 grs.
11. Reduce .83⅓ lb. apoth. to units of an ounce. *Ans.* 10 oz.
12. Reduce .16669 yd. to units of an in. *Ans.* 6 in.+
13. Reduce .259 league to miles, fur., po., and parts. *Ans.* 0 m, 6 fur. 8.64 po.

861	862	863	864	865	866	867	868	869	870
$\frac{1}{861}$	$\frac{1}{862}$	$\frac{1}{863}$	$\frac{1}{864}$	$\frac{1}{865}$	$\frac{1}{866}$	$\frac{1}{867}$	$\frac{1}{868}$	$\frac{1}{869}$	$\frac{1}{870}$

14. Reduce .7625 m. to fur. and po. *Ans.* 6 fur. 4 po.
15. Reduce .875 yd. to qrs. and na. *Ans.* 3 qrs. 2 na.
16. Reduce .9 E. English to qrs. and na. *Ans.* 4 qrs. 2 na.
17. Reduce .3375 acre to roods and sq. po. *Ans.* 1 R. 14 po.
18. Reduce .67 rood to poles and tenths. *Ans.* 26.8 po.
19. Reduce .25 hhds. to gallons and quarts. *Ans.* 15 gals. 3 qts.
20. Reduce .875 gal. to pints. *Ans.* 7 pints.
21. Reduce .8125 peck to qts. and pints. *Ans.* 6 qts. 1 pt.
22. Reduce .375 bushel to pecks and quarts. *Ans.* 1 pe. 4 qts.
23. Reduce .625 day to hours. *Ans.* 15 hrs.
24. Reduce .4 year to days, hours, and minutes. *Ans.* 146 days 2 hrs. 24 min.

RATIOS AND PROPORTIONS.

Ratio is the numerical relation of antecedent and consequent, or the number of times, or parts of times, which the former may be applied to the latter.

Thus, the ratio of 4 to 8 is 2; because 2 is the number of times which 4 may be applied to 8.

And the ratio of 8 to 4 is $\frac{1}{2}$; because 8 may be applied to 4, half of one time: and this ratio is the reciprocal of the other.

The ratio of any two numbers multiplied into a third produces a fourth proportional: thus, the ratio of 4 to 8 multiplied into 3 equals 6; therefore, 4 : 8 : : 3 : 6.

Proportions are measured distances upon the line of the prime series; because 4 measures 8 as often as 3 measures 6; and 8 applies to 4 as often as 6 applies to 3.

871	872	873	874	875	876	877	878	879	880
$\frac{1}{871}$	$\frac{1}{872}$	$\frac{1}{873}$	$\frac{1}{874}$	$\frac{1}{875}$	$\frac{1}{876}$	$\frac{1}{877}$	$\frac{1}{878}$	$\frac{1}{879}$	$\frac{1}{880}$

Numbers have to each other the same ratio as their equimultiples; that is, 4 is to 8 as 2×4 is to 2×8, or as 3×4 is to 3×8, or, as any number of times 4 is to the same number of times 8: the same is true of the equal parts.

And since the ratio is all we require, it is more easily found in low than in high numbers. It is therefore necessary to have a rule at hand for finding a common measure for high numbers, in order to obtain proportional numbers as near as possible to the first point of the series. This is called canceling.

At page 30 it was stated that 2 is a common measure of every 2d term, 3 of every 3d term, &c., in the line of numbers. It may, however, be necessary, in many cases, to use the following rule:

To find the greatest common measure of two given numbers.

Rule. Divide the greater by the less, the divisor by the remainder, this divisor also by its remainder, and so on until nothing is left; the number which divides and leaves nothing will be the greatest common measure of the given terms: for if a remainder occur that will measure the divisor, it will certainly measure the dividend which is a multiple of the divisor+itself.

The class will be able to give oral responses in most of the following examples; and where that is difficult, the slate may be used in accordance with the rule.

What is the greatest common measure

of 3 and 9?	of 16 and 24?	of $18\frac{3}{4}$ and 25?
of 6 and 9?	of 15 and 20?	of $37\frac{1}{2}$ and 100?
of 4 and 8?	of 18 and 27?	of 25 and $62\frac{1}{2}$?
of 4 and 12?	of 17 and 34?	of $16\frac{2}{3}$ and $83\frac{1}{3}$?
of 6 and 12?	of 14 and 28?	of $16\frac{2}{3}$ and 75?
of 6 and 15?	of 28 and 35?	of 27 and 72?
of 7 and 14?	of 24 and 36?	of 56 and 63?
of 5 and 15?	of 28 and 42?	of 8 and 96?
of 8 and 16?	of 28 and 36?	of $12\frac{1}{2}$ and $87\frac{1}{2}$?
of 9 and 18?	of 30 and 50?	of $22\frac{1}{2}$ and $67\frac{1}{2}$?
of 12 and 18?	of 26 and 52?	of 75 and 225?

881	882	883	884	885	886	887	888	889	890
$\frac{1}{881}$	$\frac{1}{882}$	$\frac{1}{883}$	$\frac{1}{884}$	$\frac{1}{885}$	$\frac{1}{886}$	$\frac{1}{887}$	$\frac{1}{888}$	$\frac{1}{889}$	$\frac{1}{890}$

And finding the common measure, in each case above, the class will proceed to divide each of the couplets by its proper measure, showing the ratio in its lowest numbers.

The *extreme* terms of four proportionals are the first and fourth; the *mean* terms are the second and third.

The antecedent is the *first* and the consequent is the 2d term of a ratio.

If four terms be proportionals, as 4, 8, 3, 6, they shall also be proportionals under the following changes, viz.:

1. Invert the whole order, 6 : 3 : : 8 : 4, or
only that of the couplets, 8 : 4 : : 6 : 3.
2. Transpose the couplets, 3 : 6 : : 4 : 8, or
only the means, 4 : 3 : : 8 : 6.

In any set of four proportionals, the sum or difference of the terms of the first couplet is to one of them, as the sum or difference of the terms of the second couplet is to one of these. Let 6, 3, 4, 2, be four proportionals; then 6+3 : 6, or 3 : : 4+2 : 4, or 2; and 6—3 : 6, or 3 : : 4—2 : 4 or 2.

The ratios of the sums, differences, products and quotients of the corresponding terms of different sets of four proportionals will be equal. Let 6, 12, 4, 8, be a set.

1. Divide the terms by 2; it quotes 3 : 6 : : 2 : 4.
2. Multiply these by 4; it produces 12 : 24 : : 8 : 16.
3. Subtract 1st from 2d; it leaves 9 : 18 : : 6 : 12.
4. Add 1st 2d and 3d; it amounts to 24 : 48 : : 16 : 32.

In any number of terms, as 3, 6, 12, 24, 9, 18, 48, and as many others, 2, 4, 8, 16, 6, 12, 32, which, taken two and two in order, have the same ratio, the first is to the last or other distant term of one rank, as the first is to the last or other equidistant term of the other rank: thus 3 : 48 : : 2 : 32, and 3 : 18 : : 2 : 12, &c.

A compound ratio is the product of two or more simple ratios; namely, the product of the antecedents for an antecedent, and the product of the consequents for a consequent.

The greatest advantage of canceling is seen in operations of this kind; where, among several antecedents and

891	892	893	894	895	896	897	898	899	900
$\frac{1}{891}$	$\frac{1}{892}$	$\frac{1}{893}$	$\frac{1}{894}$	$\frac{1}{895}$	$\frac{1}{896}$	$\frac{1}{897}$	$\frac{1}{898}$	$\frac{1}{899}$	$\frac{1}{900}$

as many consequents, it may happen that these, or some of them, being equal, or in the ratio of 1 to 1, may be omitted; or having a common measure, two and two, may be reduced to low numbers.

The squares and cubes of proportionals are also proportionals: thus,

If 2 : 4 : : 3 : 6; then, 4 : 16 : : 9 : 36, the squares; and if 2 : 4 : : 3 : 6; then, 8 : 64 : : 27 : 216, the cubes.

If the first be to the second as the second to the third, the last of the three is called a third proportional, and the second a mean proportional to the other two. Thus, if 2 is to 4 as 4 is to 8, 8 is a third and 4 is a mean proportional.

Direct proportion is that in which both the consequents are greater, or both less, than the antecedents.

Inverse proportion is that in which one of the consequents is less, and the other greater, than its antecedent; but in such degree, that by inverting one of the couplets the proportion will be direct.

Direct proportion makes the product of the extremes equal to the product of the means; thus, $4\times6=3\times8$, which comes from the statement, 4 : 8 : : 3 : 6.

Inverse proportion makes the product of the first couplet equal to that of the second; thus, $4\times6=3\times8$; but this comes from the statement, 4 . 6 : : 3 . 8, which may mean that 4 days of 6 hours each equal 3 days of 8 hours each; or 4 feet wide of 6 feet long equal 3 feet wide of 8 feet long; or any similar application. It would be improper to use the colon (:) in the sense *is to* in the inverse statement. The point (.) is equivalent to the oblique cross.

Rule. To find a fourth proportional to three given terms: multiply the 2d and 3d together, and divide the product by the first; the quotient will be the thing required.

Or, the product of the third term multiplied by the ratio of the first and second; or, the 2d multiplied by the ratio of the 1st and 3d, will be a fourth proportional.

901	902	903	904	905	906	907	908	909	910
$\frac{1}{901}$	$\frac{1}{902}$	$\frac{1}{903}$	$\frac{1}{904}$	$\frac{1}{905}$	$\frac{1}{906}$	$\frac{1}{907}$	$\frac{1}{908}$	$\frac{1}{909}$	$\frac{1}{910}$

Examples for the Slate.

Find a fourth proportional to 3, 6, and 12.

12 The ratio of 3 to 6 is 2. : 2×12=24.
6 or the ratio of 3 to 12 is 4. : 4× 6=24.

3)72

Ans. 24, for 3 : 6 : : 12 : 24.

Find a fourth proportional to each of the three terms here following, or find the number which should occupy the place of x.

To 4, 6, and 8,—x.
To 3, 9, and 9,—x.
To 6, 9, and 12,—x.
To 4, 8, and 12,—x.
To 4, 12, and 16,—x.
To 6, 12, and 24,—x.
To 6, 15, and 18,—x.
To 7, 14, and 21,—x.
To 5, 15, and 15,—x.
To 8, 16, and 24,—x.
To 9, 18, and 27,—x.
To 12, 18, and 36,—x.
To 11, 22, and 44,—x.
To 14, 21, and 42,—x.
To 16, 24, and 32,—x.
To 15, 20, and 45,—x.
To 18, 27, and 54,—x.
To 17, 34, and 51,—x.
To $22\frac{1}{2}$, $67\frac{1}{2}$, and 45,—x.
To 14, 28, and 28,—x.
To 28, 35, and 35,—x.
To 24, 36, and 36,—x.
To 28, 42, and 42,—x.
To 28, 36, and 56,—x.
To 30, 50, and 75,—x.
To 26, 52, and 78,—x.
To $7\frac{1}{4}$, $43\frac{1}{2}$, and $14\frac{1}{2}$,—x.
To $12\frac{1}{2}$, 15, and 25,—x.
To $18\frac{3}{4}$, 25, and $37\frac{1}{2}$,—x.
To $37\frac{1}{2}$, 100, and 75,—x.
To 25, $62\frac{1}{2}$, and 75,—x.
To $16\frac{2}{3}$, $83\frac{1}{3}$, and $8\frac{1}{3}$,—x.
To $16\frac{2}{3}$, 75, and $33\frac{1}{3}$,—x.
To 27, 72, and 81,—x.
To 56, 63, and 28,—x.
To 8, 96, and 4,—x.
To $12\frac{1}{2}$, $87\frac{1}{2}$, and $6\frac{1}{4}$,—x.
To 75, 225, and $37\frac{1}{2}$,—x.

Whenever the 2d and 3d terms happen to be the same number, the required term will be a third proportional: but the rule for finding it is similar to the rule above.

The ancient Rule of Three, which applies so much to human affairs, is similar to the rule above, for finding a 4th proportional to three given terms.

It takes hold of all the transactions of commerce, in the language of the business people, speaking after the

911	912	913	914	915	916	917	918	919	920
$\frac{1}{911}$	$\frac{1}{912}$	$\frac{1}{913}$	$\frac{1}{914}$	$\frac{1}{915}$	$\frac{1}{916}$	$\frac{1}{917}$	$\frac{1}{918}$	$\frac{1}{919}$	$\frac{1}{920}$

following manner: "If $10\frac{1}{2}$ dollars buy 3 pairs of boots, how many pairs will $35 purchase?" In this order the question is stated, and it speaks the language which all understand.

```
dols.     prs.     dols.    prs.
10.5  :   3   : :  35.0  :  10
                      3
                   ----
         10.5)105.0(10 prs. Ans.
              105
              ---
               00
```

RULE OF THREE DIRECT.

Because the Rule of Three is the application of the doctrines of proportion to business, the terms convey the idea of goods; and two of the given terms express the same kind of goods, and often in different measures, or weights, or units of unequal value; as, yds. qrs. na.; lbs. oz. drs.; bu. pe. qts.; &c. In such case, the two similar quantities must be reduced to units of the same grade: when this is done, there is no more notice to be taken of the species of goods which they represented. The two numbers are now terms of a ratio, antecedent and consequent; and if they have a common measure, may be reduced to lower terms. Neither does it make any difference whether the third given term be placed between them, or on the right of them; or whether the third multiply the second term, or the second multiply the third: it is the species of goods which the third *given* term represents that is multiplied; and it is that species which appears in the product, and also in the quotient, after dividing by the first term, or antecedent.

921	922	923	924	925	926	927	928	929	930
$\frac{1}{921}$	$\frac{1}{922}$	$\frac{1}{923}$	$\frac{1}{924}$	$\frac{1}{925}$	$\frac{1}{926}$	$\frac{1}{927}$	$\frac{1}{928}$	$\frac{1}{929}$	$\frac{1}{930}$

Proof. Invert the order of the four terms, or only of the couplets, or transpose the couplets; by these three changes each of the given terms becomes the 4th term, and will result from a similar operation.

Examples for the Slate.

1. If 8 yards of cloth cost $28, what will 96 yards of the same kind come to?

```
   yds. yds.    $    $
As 8 : 96 : : 28 : x.
               96
              ---
              168
             252
            -----
           8)2688
            -----
      Ans. $336
```

The operation is here presented in full, without contraction; but since 8 is a common measure of 8 and 96, and the reduced terms are 1 and 12, therefore,

1 : 12 : : 28 : 336, *Ans.*

2. What will 76 yards 2 qrs. come to, if 9 yards cost 22 dols. 50 cents?

```
   yds. dols.    yds.
As 9 : 22.5 : : 76.5 : x.
         76.5
        ------          4)2.0 qrs.
         1125             ---
        1350            76.5 yds.
       1575
      --------
     9]1721.25
      --------
Ans. $191.25
```

Here, also, the operation is given in full; but as 9 is a com. meas. for 9 and 22.5, the terms reduced would be 1 and 2.5; that is, 1 yd. costs 2.5$, and $2.5 \times 76.5 =$ 191.25 dols.

Class Exercise upon the examples here annexed.

How many examples, or questions, in the following collection?

In example 1, *what kind* of goods, or things, is twice mentioned? In example 2, what kind of goods is twice mentioned? In example 3, the same? also in example 4? and so on to the last.

In example 1, *how much* goods in each of the two similar quantities? In example 2, *how much* goods in each of the two similar

931	932	933	934	935	936	937	938	939	940
$\frac{1}{931}$	$\frac{1}{932}$	$\frac{1}{933}$	$\frac{1}{934}$	$\frac{1}{935}$	$\frac{1}{936}$	$\frac{1}{937}$	$\frac{1}{938}$	$\frac{1}{939}$	$\frac{1}{940}$

quantities? In example 3, the same? also in example 4? and so on to the last.

In what examples are the goods measured? In what examples are the goods weighed? In what examples are the goods only counted? Are there several names of measures in any example? Are there several names of weights in any example? Which are the rules for reducing all the names of measures or weights into one name? Would you prefer to reduce them from high to low, or from low to high grades? What direction to reduction does multiplication give?—does division give?

3. The salary of a clerk is $1000 per annum; how much, then, is his pay for 1 calendar month? also for 1 day?

c.m. dols. c.m. dols.
12 : 1000 : : 1 : $83\frac{1}{3}$. *Ans.*

$83\frac{4}{12}$

ds. dols. d. dols.
365 : 1000 : : 1 : $2.7397\frac{19}{73}$.
730

2700
2555

1450
1095

3550
3285

2650
2555

rem. 95 $\frac{95}{365}=\frac{19}{73}$.

There is no multiplier in either of these examples, therefore the quotient is placed in the line of proportionals, and the divisor under the last remainder.

The fraction $\frac{4}{12}=\frac{1}{3}$, dividing the terms by 4. $\frac{95}{365}=\frac{19}{73}$, dividing the terms by 5, the common measure.

4. The cost of 7 cwt. 1 qr. of sugar being $50.75, how much, at the same rate, will 43 cwt. 2 qrs. amount to?

5. The cost of 43.5 cwt. of cheese being $$304\frac{1}{2}$, how much of the same kind will 50 dols. 75 cts. purchase?

6. If 7.25 cwt. of bacon cost $$50\frac{3}{4}$, how much of the same quality will $304.5 purchase?

7. If 304 dols. 50 cents buy 43.5 cwt. of best white soap, what is the cost of 7 cwt. 1 qr. of the same kind?

8. If 24 dols. 36 cents buy 12 yards of cloth, how many yards will $292.32 purchase?

941	942	943	944	945	946	947	948	949	950
$\frac{1}{941}$	$\frac{1}{942}$	$\frac{1}{943}$	$\frac{1}{944}$	$\frac{1}{945}$	$\frac{1}{946}$	$\frac{1}{947}$	$\frac{1}{948}$	$\frac{1}{949}$	$\frac{1}{950}$

9. If $292.32 purchase 144 pairs of boots, how many pairs will $24.36 purchase?

10. 144 coarse shawls cost 292 dols. 32 cts., what will 12 shawls of the same kind come to?

11. If 12 wooden clocks cost 24 dols. 36 cts. how much will 144 clocks of the same kind amount to?

12. If 14 lbs. of fine tea cost 14 dols. 33 c. 7 m., how many lbs. of the same quality will $200.718 purchase?

13. The cost of 196 pairs of silk stockings being $200.718, how many such pairs will $14.337 purchase?

14. $14.337 being paid for 14 silver spoons, how much will 196 spoons of the same kind amount to?

15. 196 bushels of wheat cost $200.718, how much will 14 bushels of the same kind come to?

16. If 19 dols. 44 cts. buy 18 yards of merino plaid, how many yards will $349.92 purchase?

17. If 324 days' wages amount to 349 dols. 92 cts., how many days' labor will $19.44 purchase?

18. If 349 dols. 92 cts. purchase 324 boys' caps, how much will eighteen caps of the same kind come to?

19. If $19.44 purchase 18 girls' bonnets, how much will 324 bonnets of the same kind come to?

20. If $12.912 purchase 10 pairs of shoes, how many pairs may be bought for $129.12?

21. If $129.12 buy 100 acres of public land, how many acres can one purchase for $12.912?

22. If 100 bushels of peaches cost $129.12, how much will 10 bushels of the same kind come to?

23. If 10 turkeys cost $12.912, how much will a flock of 100 such turkeys amount to?

24. If $56.566 buy 16 cords of wood, how many cords will $905.056 purchase?

25. If $905.056 purchase 256 tons of coal at the mine, how many tons will $56.566 purchase?

26. If 16 best silk hats cost $56.566, how much will 256 such hats amount to?

27. If 256 fat sheep cost $905.056, how much will 16 such sheep amount to?

951	952	953	954	955	956	957	958	959	960
$\frac{1}{951}$	$\frac{1}{952}$	$\frac{1}{953}$	$\frac{1}{954}$	$\frac{1}{955}$	$\frac{1}{956}$	$\frac{1}{957}$	$\frac{1}{958}$	$\frac{1}{959}$	$\frac{1}{960}$

28. A gentleman saves $1200 per annum of his income, which is 3858.24 dollars; what are his daily expenses? *Ans.* 7.2828 dols.

29. A draper bought 4 pieces of cassimer containing 100½ yards, at 1.6 dollars per yard; what was the amount? *Ans.* $160.8

30. If 90 lbs. of tea cost $47.7, how much did 1 lb. cost? how much did 74 lbs. cost? how much did 7 chests each 74 lbs. cost?

Ans. 1 lb. cost 53 cents. Seek the others.

31. Tea at 50 cts., 75 cts., 1 dol., and 1 dol. 12½ cts. per lb.; what equal quantity of each kind will 675 dollars purchase? *Ans.* 200 lbs. of each.

32. The price of standard silver (9 parts silver and 1 part copper) is 116$\frac{1}{4}\frac{2}{3}$ cts. per ounce; then what quantity may be bought for $100? *Ans.* 7 lbs. 2 oz. troy.

33. The clothing of a regiment of 750 men costs $13875; then how much will the clothing of an army of 50000 men cost the United States?

Ans. 925000 dollars.

34. What will be the expense of keeping 20 horses for a year, at 37½ cents per day for each? *Ans.* $2737.5.

35. If an estate be worth $1539.2, and the tax assessed at 3½ cts. upon the dollar; what is the remainder worth per annum? *Ans.* $1485.328.

36. If 300 lbs. of tea be sold for 255 dols. making the gain 11 dols. 25 cts, what was the prime cost per lb.?

Ans. 81¼ cents.

37. If 300 lbs. of tea cost 255 dols., how must it be sold per lb. to make the whole gain 45 dollars?

Ans. 1 dol. per lb.

38. A butcher laid out in oxen $1151.5, at the rate of $245 for every 10 oxen; how many oxen did the sum purchase? *Ans.* 47 oxen.

39. What will 4 pieces of linen come to, containing 23, 24, 25 and 27 yards, at $7 for 16 yards?

Ans. $43.31¼.

40. The Eagle or $10 piece (new coin) weighs 10 dwts. 18 grs. of standard gold, and is worth 93 cts. per

961	962	963	964	965	966	967	968	969	970
$\frac{1}{961}$	$\frac{1}{962}$	$\frac{1}{963}$	$\frac{1}{964}$	$\frac{1}{965}$	$\frac{1}{966}$	$\frac{1}{967}$	$\frac{1}{968}$	$\frac{1}{969}$	$\frac{1}{970}$

dwt., the British sovereign weighs 5 dwts. 3 grs. and is worth 94$\frac{3}{4}$ cts. per dwt.; how many dollars are equal to 32 of each? *Ans.* $475.39.

41. If 802 cwt. 1 qr. 14 lbs. of hay be sold at the rate of 18\frac{3}{4}$ per ton, how much will the whole amount to? *Ans.* $752.2265625.

42. If 660 lbs. of coffee cost $63.80 cts., how much is that per. lb.? *Ans.* 9$\frac{2}{3}$ cents.

43. A merchant laid in store 2 casks of oil containing 73 gals. 2 qts. each, 2 casks containing 87 gals. 3 qts. each, and 2 casks containing 68 gals. 2 qts. each, for which he paid at the rate of $11 for 16 gallons; how much is the entire cost? *Ans.* 315.9\frac{1}{16}$.

44. A southern merchant received from the north 108 pairs of boots, at 2.16\frac{2}{3}$ cts. per pair; the cost of transportation was $6: how shall he sell them by the pair to make 20 cts. on the dollar? *Ans.* 2.66\frac{2}{3}$.

SINGLE RULE OF THREE INVERSE.

If there be two equal rectangles, as **BG, DH** (see the figures in the margin), their sides, taken reciprocally, are proportionals; thus, AB : CD : : CH : AG. Here the length of one figure is to the breadth of the other, as the length of the latter is to the breadth of the former. But if we should say that the length is to the breadth of one figure as the length is to the breadth of the other, the proportion would be inverted; thus, AB×AG=CD×CH.

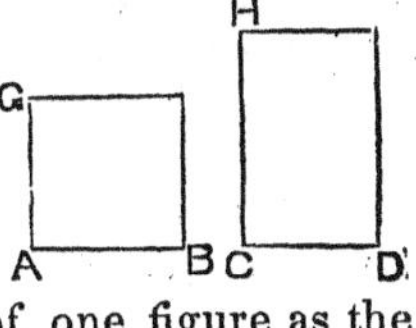

The rectangles 3×8=4×6, 4×6=2×12, taken as above, are examples of proportions inverted; and the same relations are seen all along the lines of numbers and reciprocals at the heads of these pages

971	972	973	974	975	976	977	978	979	980
$\frac{1}{971}$	$\frac{1}{972}$	$\frac{1}{973}$	$\frac{1}{974}$	$\frac{1}{975}$	$\frac{1}{976}$	$\frac{1}{977}$	$\frac{1}{978}$	$\frac{1}{979}$	$\frac{1}{980}$

For, directly, and inversely,

$2 : 3 :: \frac{1}{3} : \frac{1}{2}$, or $\frac{2}{6} : \frac{3}{6}$; $2 . \frac{1}{2} :: 3 . \frac{1}{3}$.
$3 : 4 :: \frac{1}{4} : \frac{1}{3}$, or $\frac{3}{12} : \frac{4}{12}$; $3 . \frac{1}{3} :: 4 . \frac{1}{4}$.
$4 : 5 :: \frac{1}{5} : \frac{1}{4}$, or $\frac{4}{20} : \frac{5}{20}$; $4 . \frac{1}{4} :: 5 . \frac{1}{5}$.
$5 : 6 :: \frac{1}{6} : \frac{1}{5}$, or $\frac{5}{30} : \frac{6}{30}$; $5 . \frac{1}{5} :: 6 . \frac{1}{6}$.
$6 : 20 :: \frac{1}{20} : \frac{1}{6}$, or $\frac{6}{120} : \frac{20}{120}$; $6 . \frac{1}{6} :: 20 . \frac{1}{20}$.

The colon (:) in the sense of *is to* cannot be properly used between the inverted terms; but a point (.) or the oblique cross (×) to denote multiplication should be used.

When few agents can do the work of many; when one dimension supplies the defect of another; when force, or velocity, or number, accounts for a diminution of time; these and their contraries indicate proportions inverted. Now such distinctions appear, not in the numbers, but in other properties of the quantities; or, in a different extension, or modification of their parts, and can be known only from the enunciation of the question. Therefore,

If more do more, or less do less, respect,
It is a question of the Rule Direct:
But less requiring more, and greater less,
Do questions of the Inverse Rule express.

RULE. State the terms so that, if the required consequent is to be less than its antecedent, the less of the two similar given terms shall be consequent; and if the required consequent is to be greater than its antecedent, the greater of the two similar given terms shall be consequent. This makes a direct statement of the terms; and so, the product of the means divided by the given extreme will be the term required.

Contraction. When the first and second, or the first and third terms, have a common measure, divide them by it, canceling the terms divided, and using their quotients.

Examples for the Slate.

1. If 5 men can do 100 yards of a work in 24 days,

981	982	983	984	985	986	987	988	989	990
$\frac{1}{981}$	$\frac{1}{982}$	$\frac{1}{983}$	$\frac{1}{984}$	$\frac{1}{985}$	$\frac{1}{986}$	$\frac{1}{987}$	$\frac{1}{988}$	$\frac{1}{989}$	$\frac{1}{990}$

what number of men must be employed to finish an equal quantity in 15 days? or in 1 day?

Here since 15 days is less time than 24 days, the number of men must exceed 5.

Therefore 15 days : 24 days : : 5 men : 8 men.

$$\begin{array}{r} 5 \\ \hline 15)120 \text{ men will require 1 day.} \\ \hline \textit{Ans. } 8 \text{ men require 15 days.} \end{array}$$

I said mentally, 15 is to 24 as $\frac{1}{3}$ of 15 is to $\frac{1}{3}$ of 24=8.

2. What length of 12 inches wide makes a square foot? also of 9 inches wide? of 8 inches wide? of 6 inches wide? of $4\frac{1}{2}$ inches wide? of 4 inches wide? of 3 inches wide? of 2 inches wide? of 1 inch wide?

Ans. 12 in. of 12 in., 16 in. of 9 in. &c.

3. What length of yard wide carpet will cover a floor 18 ft. long and $16\frac{1}{2}$ ft. broad? also of 2 yards wide? of 3 yards wide? of 6 yards wide?

Ans. 33 of 1 yd. $16\frac{1}{2}$ of 2, &c.

4. If 6352 hewn stones of 3 feet long complete a certain wall, how many stones of 2 feet long, other dimensions being equal, will complete an equal quantity? how many stones of 4 ft. long would do it? also of 6 ft. long? and of 12 ft. long? *Ans.* 9528 of 2 ft., 4764 of 4 ft., &c.

5. A garrison of 480 men having provisions for 60 days, how long would the provisions serve 240 men? how long would they serve 120 men? how long would they serve 360 men? 720 men? 960 men? *Ans.* 120 ds., &c.

6. In 12 days of 4 hours each a certain work was finished; how many days of 6 hours? of 8 hours? of 12 hours? of 24 hours, would the same work require?

Ans. 8 ds. of 6 hrs., &c.

7. A plain or prairie pastured 2000 horses for 18 days, how many days would it have pastured 1000 horses? 500 horses? 1500 horses? 2500 horses? 3000 horses?

Ans. 36 ds., &c.

8. The governor of a besieged town has provisions for

991	992	993	994	995	996	997	998	999	1000
$\frac{1}{991}$	$\frac{1}{992}$	$\frac{1}{993}$	$\frac{1}{994}$	$\frac{1}{995}$	$\frac{1}{996}$	$\frac{1}{997}$	$\frac{1}{998}$	$\frac{1}{999}$	$\frac{1}{1000}$

54 days, the ration of bread being $1\frac{1}{2}$ lb.; but fearing that his succors will be delayed, he wishes to hold out for 81 days; how much, in that case, should the ration of bread be? *Ans.* 1 lb.

9. How many yards of stuff 3 quarters wide, will line a cloak $1\frac{3}{4}$ yards long, and $3\frac{1}{2}$ yards wide? *Ans.* 8 yds. 6 in.

10. If a footman perform a journey in 5 days of 12 hours long, how many days will he require of 10 hours? of $7\frac{1}{2}$ hours? of 6 hours? *Ans.* 6 ds. of 10 hrs., &c.

11. If I lend my friend 120 dollars for $4\frac{1}{2}$ months, how many months should he lend me 90 dollars? or 180 dollars? or 60, to requite my kindness? *Ans.* 90 dols. for 6 mos., &c.

12. A power or weight of 32 lbs. is placed 4 ft. 6 in. from the centre of motion of a steelyard, what weight will it equipoise 27 in. from the same centre at the draft end? or at 18 inches? or at 9 inches? or at 6 inches? or at 3 inches? or at $1\frac{1}{2}$ inches? *Ans.* 64 lbs. at 27 in., &c.

13. There are three windows which measure each 27 ft. 48 in. of glazing; it is required to have an equal quantity in 1 window, or in two windows; how much is each window in these cases to be?

14. Let a room measure 24 feet long and $16\frac{1}{2}$ feet wide, then, if another room be equal to it, and only 22 ft. long, what should be the width of the latter? also, if another room be 20 feet long, how wide should it be to equal the first? *Ans.* 18 ft. by 22, &c.

Promiscuous Examples.

1. Bought four casks of raisins, weighing each 3 cwt. 1 qr. 7 lbs., at $6.25 per cwt.; what is the amount? *Ans.* 82 dols. $81\frac{1}{4}$ cts.

2. Six men build a house in 9 months, how many months would 9 men require? how many months would 18 men require? how many would 27 men require? *Ans.* 6 mos. of 9 men, &c.

3. A grocer bought 2 qrs. 14 lbs. of cloves at 75 cts.

1001	1002	1003	1004	1005	1006	1007	1008	1009	1010
$\frac{1}{1001}$	$\frac{1}{1002}$	$\frac{1}{1003}$	$\frac{1}{1004}$	$\frac{1}{1005}$	$\frac{1}{1006}$	$\frac{1}{1007}$	$\frac{1}{1008}$	$\frac{1}{1009}$	$\frac{1}{1010}$

per lb., and sold them at the rate of $89.6 per cwt.; how much did he gain or lose by this trade?

Ans. gained $3.5.

4. If 10 men mow a meadow in 3 days, how many days would 6 men require? how many days of 5 men? how many of 1 man? how many of 8 men?

Ans. 5 ds. of 6 men, &c.

5. If from a rule of three feet long,
The shadow five is made;
What is that steeple's height in feet,
That is ninety yards in shade? *Ans.* 162 ft.

6. What length, of 8 poles wide will make an acre? of 4 poles wide? of 2 poles wide? of 1 pole wide? of 20 poles wide; of 40 poles wide? of 80 poles wide? of 160 poles wide? *Ans.* 20 po. by 8, &c.

7. If the weight of an eagle be 10 dwts. 18 grs. of gold, how much will be the weight of 180 half eagles?

Ans. 4 lbs. 7 dwts. 12 grs.

8. If the small loaf weigh 20 ounces when wheat is $1.20 per bushel, how much should it weigh when wheat is $1.50 per bushel? also, when wheat is $1 per bushel?

Ans. 16 oz. at $1.50, &c.

9. The avoirdupois pound weighs 7000 grains, and the troy pound 5760 grains; then how much of the latter is equal to one cwt. of the former? *Ans.* $136\frac{1}{9}$ lb. troy.

10. If the bushel measure $2150\frac{2}{5}$ cubic inches, how many bushels of wheat will the toll chest contain which measures $31.73\frac{1}{3}$ cubic feet? *Ans.* $25\frac{1}{2}$ bu.

11. A certain building was raised in 8 months by 120 workmen, but being demolished, it is to be rebuilt in 2 months; how many men must be employed upon it?

Ans. 480 men.

12. How much will 4 dozen and 8 pairs of shoes amount to at $$1\frac{1}{8}$ per pair? *Ans.* 63 dols.

13. If the transportation of $7\frac{1}{2}$ cwt. for 150 miles cost $12, how far should 3 tons be carried for the same money? *Ans.* 18 m. 6 fur.

14. If 2 qrs. 22 lbs. of sugar cost $6.24, how much is that for 1 cwt., or 112 lbs.? *Ans.* $8.96.

1011	1012	1013	1014	1015	1016	1017	1018	1019	1020
$\frac{1}{1011}$	$\frac{1}{1012}$	$\frac{1}{1013}$	$\frac{1}{1014}$	$\frac{1}{1015}$	$\frac{1}{1016}$	$\frac{1}{1017}$	$\frac{1}{1018}$	$\frac{1}{1019}$	$\frac{1}{1020}$

15. At $2\frac{1}{4}$ dollars per 100, what will 9000 feet of plank amount to? *Ans.* 202.5 dols.

16. Laid out $36 in calico and linen; the value of the calico was 15 dollars, and the quantity of linen 70 yards; and there were 7 yards of linen to 5 yards of calico; what was the price per yard? also how many yards of calico? *Ans.* price 30 cts. and calico 50 yds.

17. A cistern is emptied by a vent in 12 hours; how many vents of the same capacity would fill it in 6 hours? in 4 hours? in 3 hours? in 2 hours? in 1 hour? in .5 of an hour?

18. Bought 80 yards of cloth for 432 dollars, which being damaged, am willing to lose $20 on the whole: how should it then be sold per yard? *Ans.* $5 15 cts.

19. If 100 dollars in 12 months gain $5, how much will the same sum gain in 6 months? in 3 months? in 1 month? in 8 months? in 16 months? in 18 months? in 21 months? *Ans.* $$2\frac{1}{2}$ in 6 mos., &c.

20. If 150 dollars in 15 months gain 9 dols. 37 cts. 5 m. interest, how much will the same sum gain in 12 months? *Ans.* $7.50.

21. Bought three tuns of oil for 409.31 dollars, 85 gallons of which having leaked out, how should the residue be sold per gallon to sustain this loss?

Ans. at 61 cents.

22. The shadow of a staff of 4 feet long, at a certain hour, is 7 feet, what then is the height of a steeple, whose shade at the same time is 198 feet?

Ans. $113\frac{1}{7}$ feet.

23. What sum at interest for 1 month would gain the same that 240 dollars would gain in 12 months? what sum also at 2 months? what sum also at 6 months?

Ans. $2880 for 1 mo. &c.

24. If 8 boarders consume a barrel of flour in 14 days, how many days should it serve 4 boarders? how many days should it serve 2 boarders? also, how many days should it serve 10? 12?

25. A perpendicular pole of 50 ft. 11 in. casts a shadow of 98 feet 6 inches, when the sun is vertical; then how

1021	1022	1023	1024	1025	1026	1027	1028	1029	1030
$\frac{1}{1021}$	$\frac{1}{1022}$	$\frac{1}{1023}$	$\frac{1}{1024}$	$\frac{1}{1025}$	$\frac{1}{1026}$	$\frac{1}{1027}$	$\frac{1}{1028}$	$\frac{1}{1029}$	$\frac{1}{1030}$

broad is the river, which runs due west, on the north side of a steeple 300 ft. 8 in. high, 20 ft. 6 in. of the shadow falling between the base and the brink of the water, and 30 ft. 9 in. being projected on the opposite side?

Ans. 176 yds. 2 ft. 4 in.

DOUBLE RULE OF THREE.

This rule, which is sometimes called Compound Proportion, because a ratio is used equal to the product of two or more ratios, proposes an odd number of terms, not less than five: of these terms the simple majority are antecedents; the rest are consequents; and one consequent is required.

The terms are of similar kinds, two and two; and the design of the work is to find a consequent to correspond with one of the antecedents, which is unmatched among the given terms.

Set that term the quality of which is sought on the right; arrange the other given terms two and two, in the manner of a ratio, with the colon between them on the left; observing that if upon comparison with each couplet, the required consequent is to be greater than its antecedent, the greater term of such couplet is to be the consequent; but if the required consequent is to be less than its antecedent, the less term of such couplet is to be the consequent.

Reduce the terms of each couplet to units of the same grade, and any antecedent and consequent which have a common measure to their lowest terms. Then multiply all the antecedents, except the odd one, for an antecedent, and all the consequents for a consequent: This antecedent *is to* its consequent *as* the odd antecedent *is to* the consequent required; which is found like any other fourth proportional.

1031	1032	1033	1034	1035	1036	1037	1038	1039	1040
$\frac{1}{1031}$	$\frac{1}{1032}$	$\frac{1}{1033}$	$\frac{1}{1034}$	$\frac{1}{1035}$	$\frac{1}{1036}$	$\frac{1}{1037}$	$\frac{1}{1038}$	$\frac{1}{1039}$	$\frac{1}{1040}$

Examples for the Slate.

1. If 3 men in 4 days reap 12 acres, how many acres will 6 men reap in 12 days?

```
m. 3 :  6 m.
ds. 4 : 12 ds.
    —    —      A.   A.
   12 : 72 : : 12 : 72
               12
              ——
          12)864
              ——
      Ans. 72 A.
```

Multiplying the antecedents for an antecedent, and the consequents for a consequent, I said mentally, 12, the compound antecedent, *is to* 72, its consequent, *as* 12, the given acres, *is to* 72, which is the number of acres required.

2. If 100 dollars in two years gain 10 dollars, what should be the interest of 750 dollars for 12 years? *Ans.* 450 dols.

3. If 12 oxen in 16 days consume 1 ton 14 cwt. 1 qr. 4 lbs. of hay, what quantity will 24 oxen consume in 48 days? and 1 ox in 1 day? *Ans.* 10 T. 5 cwt. 2 qrs. 24 lbs.

4. If 8 cwt. may be transported 128 miles for 21 dollars, how much should the freight of 16 cwt. 64 miles amount to? *Ans.* 21 dollars.

5. If 12 men in 8 days gain 90 dollars, how much should 20 men gain in 15 days? *Ans.* 281.25 dols.

6. If 600 dollars in 60 days gain 6 dollars, how much should 240 dollars gain in 480 days? *Ans.* 19 dols. 20 c.

7. If 200 lbs. be conveyed 40 miles for $2, what should the carriage of 20200 lbs. 60 miles come to? *Ans.* 303 dols.

8. If a family of 8 persons expend 800 dollars in 9 months, how much should be the expenses of 18 persons for 12 months? *Ans.* 2400 dols.

9. The hire of 4 men for 14 days being $64, how much should 14 men receive for 10 days? *Ans.* 160 dols.

10. If 6 men mow 72 acres in 4 days, how many acres should 8 men mow in 10 days? *Ans.* 240 acres.

1041	1042	1043	1044	1045	1046	1047	1048	1049	1050
$\frac{1}{1041}$	$\frac{1}{1042}$	$\frac{1}{1043}$	$\frac{1}{1044}$	$\frac{1}{1045}$	$\frac{1}{1046}$	$\frac{1}{1047}$	$\frac{1}{1048}$	$\frac{1}{1049}$	$\frac{1}{1050}$

DOUBLE RULE OF THREE INVERSE.

The rule already given is of general application, viz. :

If the required consequent, in view of each condition of the question, should be *greater* or *less* than its antecedent, make the *greater* or *less* of the similar given terms the consequent, as the case may be.

Examples for the Slate.

1. If 6 men work 4 days for £36, how many days should 12 men work for $54?

12 : 6 is the ratio of the times in which the men should work ; and 36 : 54 is the ratio of the times for which the money will pay. Therefore,

```
 m. 12  : m. 6,  or 2 : 1
    36$ :   54$, or 2 : 3
    ---    ---      -   -       ds. ds.
   432  :  324  : : 4 : 3 : : 4 : 3 Ans.
             4
          ----
     432)1296(3 days, Ans.
         1296
```

2. The carriage of 200 lbs. for 40 miles being $2, how many lbs. may be conveyed 60 miles for $303?

Ans. 20200 lbs.

3. If 27 dollars be the hire of 9 men for 3 days, how many days should 24 men work for $144?

Ans. 6 days.

4. If 2.4£ be the hire of 4 men for 3 days' labor, how many men will £9 12s. compensate for 8 days' labor?

Ans. 6 men.

5. If 8 boarders consume 72 lbs. of animal food in 12 days, in what time will a family of 12 persons consume 216 lbs.? *Ans.* 24 days.

6. If 120 men are employed 8 days in quarrying 75

1051	1052	1053	1054	1055	1056	1057	1058	1059	1060
$\frac{1}{1051}$	$\frac{1}{1052}$	$\frac{1}{1053}$	$\frac{1}{1054}$	$\frac{1}{1055}$	$\frac{1}{1056}$	$\frac{1}{1057}$	$\frac{1}{1058}$	$\frac{1}{1059}$	$\frac{1}{1060}$

rods of stone, how many days will 80 men require to quarry 300 rods? *Ans.* 48 days.

7. If 48 men in 9 days excavate 80 perches of canal, in how many days will 100 men excavate 200 perches? *Ans.* $10\frac{4}{5}$ days.

8. If $100 principal in 12 months gain $6, what principal will gain $12 in 8 months? *Ans.* $300.

9. If 20 acres of grass feed 12 oxen 16 days, how many days will 30 acres pasture 18 oxen? *Ans.* 16 days.

10. If 16 men mow 80 acres of meadow in $2\frac{1}{2}$ days, how many men should be employed to mow 120 acres in 4 days? *Ans.* 15 men.

Promiscuous Examples.

1. Let 50 dollars at interest for 8 months amount to 52 dollars; how much is the rate per cent. per annum? *Ans.* 6 per ct. per an.

2. The carriage of 24 cwt. 48 miles is 6 dollars; what distance then, should 72 cwt. be carried for 9 dollars? *Ans.* 24 miles.

3. If 150 dollars in 15 months gain $$9\frac{3}{8}$, how much will $100 gain in 12 months? *Ans.* $5.

4. The interest of $100 in 12 months being $5, how much will $150 gain in 15 months? *Ans.* $9.375.

5. If in 12 months $100 gain $5, what sum at interest 15 months will gain $9.375? *Ans.* $150.

6. If $9.375 be gained in 15 months by $150, what principal at interest 12 months will gain $5? *Ans.* $100.

7. In what time will $100 gain $5, if 150 dols. gain $9.375 in 15 months? *Ans.* 12 months.

8. The interest of $100 in 12 months is $5, in what time will $150 gain $9.375? *Ans.* 15 months.

9. If 3 men can plow 12 acres in 24 hours, in what time should 6 men plow 20 acres? *Ans.* 20 hours.

10. If 4 men subsist 80 days upon $160, how long may 8 men subsist upon $480? *Ans.* 120 days.

11. If the interest of $50 for 8 months be $2, what sum at interest 4 months will gain $4? *Ans.* $200.

1061	1062	1063	1064	1065	1066	1067	1068	1069	1070
$\frac{1}{1061}$	$\frac{1}{1062}$	$\frac{1}{1063}$	$\frac{1}{1064}$	$\frac{1}{1065}$	$\frac{1}{1066}$	$\frac{1}{1067}$	$\frac{1}{1068}$	$\frac{1}{1069}$	$\frac{1}{1070}$

12. If 12 horses remove 72 tons in 3 days, how many horses must be employed to remove 108 tons in 9 days? *Ans.* 6 horses.

13. A western traveller performed a journey of 216 miles in 16 days of 14 hours long, then how many days of 12 hours will he require in travelling 852 miles? *Ans.* 57 days $7\frac{11}{23}$ hrs.

14. If 120 horses in 6 days consume 360 bushels of oats, how many days will 480 bushels serve 240 horses? *Ans.* 4 days.

15. The standard gold of the United States is worth 93 cents per dwt., and the new Eagle weighs 10 dwts. 18 grs.; also 16 times the same weight in silver is worth $10: how much then should be the price of 1 ounce of standard silver? *Ans.* $116\frac{12}{43}$ cents.

16. If 3000 lbs. of beef serve 340 men 15 days, what quantity will 120 men require in 25 days? *Ans.* 15.75625 cwt.

17. If a barrel of mackerel serve a family of 8 persons for 3 months, how many barrels should 16 persons use in 12 months? *Ans.* 8 barrels.

18. If 100 men, in 6 days of 10 hours long, can dig a trench 200 yds. long, 3 wide and 2 deep; in how many days of 8 hours long, will 180 men dig another trench 360 yards long, 4 wide and 3 deep? *Ans.* 15 days.

MERCANTILE PRACTICE.

The design of this rule is to find, by the shortest process, the value of any quantity of goods, having the price of a unit given.

When the given quantity is a single part, or
The complement of a single part, or
A multiple and a single part, or
A multiple and the complement of a single part of the

1071	1072	1073	1074	1075	1076	1077	1078	1079	1080
$\frac{1}{1071}$	$\frac{1}{1072}$	$\frac{1}{1073}$	$\frac{1}{1074}$	$\frac{1}{1075}$	$\frac{1}{1076}$	$\frac{1}{1077}$	$\frac{1}{1078}$	$\frac{1}{1079}$	$\frac{1}{1080}$

unit of which the price is given, the labor may be, in some instances, greatly abridged.

It is a similar case, when the price is a single part, or the complement of a single part, or either of these joined to a multiple of the money unit.

When the given quantity is not found in any of these relations to the unit of which the price is given, it is either a simple number in the denomination of the unit, or should be so reduced.

The questions on pages 64, 65, and 66, are intended as auxiliary to the modes of operation under this rule. The class will therefore write out the parts and complements there referred to, as in the following

Examples of Parts and Complements of Parts of $1.

50 cts. $=\frac{1}{2}$,	comp. 50 cts.	$12\frac{1}{2}$ cts. $=\frac{1}{8}$,	comp. $87\frac{1}{2}$ cts.
$33\frac{1}{3}$ " $=\frac{1}{3}$,	" $66\frac{2}{3}$ "	10 " $=\frac{1}{10}$,	" 90 "
25 " $=\frac{1}{4}$,	" 75 "	$8\frac{1}{3}$ " $=\frac{1}{12}$	" $91\frac{2}{3}$ "
20 " $=\frac{1}{5}$,	" 80 "	$6\frac{1}{4}$ " $=\frac{1}{16}$,	" $93\frac{3}{4}$ "
$16\frac{2}{3}$ " $=\frac{1}{6}$,	" $83\frac{1}{3}$ "	5 " $=\frac{1}{20}$,	" 95 "

Case 1. When the price is a multiple of some money unit, as $1, 1 ct., 1£, &c., and the quantity a simple number, multiply the price by the quantity, the product will be the answer, in the denomination of the money unit, which reduce, if necessary, to the money of account.

Examples for the Slate.

1. 254 cwt. of hops at $28 per cwt.?

```
 254      or thus,    254
  28                    4×7=28
 ----                 ----
2032                  1016
508                      7
 ----                 ----
$7112 Ans.           $7112 Ans.
```

2. 324 yards of cloth at 6 dollars per yard?
Ans. 1944 dols.

1081	1082	1083	1084	1085	1086	1087	1088	1089	1090
$\frac{1}{1081}$	$\frac{1}{1082}$	$\frac{1}{1083}$	$\frac{1}{1084}$	$\frac{1}{1085}$	$\frac{1}{1086}$	$\frac{1}{1087}$	$\frac{1}{1088}$	$\frac{1}{1089}$	$\frac{1}{1090}$

3. 327 lbs. of coffee at 6 cents per lb.? *Ans.* 19.62 dols.
4. 871 cwt. of iron at \$3 per cwt.? *Ans.* 2613 dols.
5. 472 cwt. of sugar at \$8 per cwt.? *Ans.* \$3776.
6. 427 bushels of wheat at 11s. per bu.? *Ans.* £234 17s.
7. 247 bushels of rye at 9s. per bu.? *Ans.* £111 3s.
8. 274 bushels of maize at 7s. per bu.? *Ans.* £95 18s.
9. 742 sheep at \$3 per head? *Ans.* \$2226.
10. 724 yds. of alpaca at 50 cts.? *Ans.* \$362.
11. 262 lbs. of sugar at 9 cts.? *Ans.* \$23.58.
12. 622 lbs. of coffee at 8 cts. per lb.? *Ans.* \$49.76.
13. 385 lbs. of Hyson tea at 8s. per lb.? *Ans.* £154.
14. 324 yds. of calico at 25 cts. per yd.? *Ans.* \$81.
15. 342 yds. of vesting at 6s. per yd.? *Ans.* £102 12s.
16. 312 days' wages at \$2 per day? *Ans.* \$624.
17. 526 yds. linen at 2s. per yard? *Ans.* £52 12s.

Note. In this last example, it appears that, when the price is 2s., the answer is found by doubling the figure occupying the units' rank: thus, 526 at 2s. is £52 12s., therefore at 4s., 6s., or any multiple of 2s., the same multiple of the price at 2s. would be the answer.

Case 2. When the price is a single part of the money unit, and the quantity a simple number; divide the quantity by the denominator of the part; the quotient will be the answer, in the denomination of the money unit, which reduce if necessary to the money of account.

Examples for the Slate.

1. 560 bushels of oats at 50 cts., and at 10s. per bu.?

cts.	560	s.	560
$50=\frac{1}{2}$	\$280 *Ans.*	$10=\frac{1}{2}$£	280£ *Ans.*

2. 255 willow baskets at $33\frac{1}{3}$ cts.? *Ans.* \$85.
3. 375 lbs. of butter at 25 cts.? *Ans.* \$93.75.
4. 546 pairs of short hose at 20 cts.? *Ans.* \$109.2.
5. 768 lbs. of candles at $16\frac{2}{3}$ cts.? *Ans.* \$128.

1091	1092	1093	1094	1095	1096	1097	1098	1099	1100
$\frac{1}{1091}$	$\frac{1}{1092}$	$\frac{1}{1093}$	$\frac{1}{1094}$	$\frac{1}{1095}$	$\frac{1}{1096}$	$\frac{1}{1097}$	$\frac{1}{1098}$	$\frac{1}{1099}$	$\frac{1}{1100}$

6. 812 yds. of calico at 12½ cts.? *Ans.* $105.25.
7. 975 lbs. of lard at 10 cts.? *Ans.* $97.5.
8. 468 lbs. sugar at 8⅓ cts.? *Ans.* $39.
9. 156 ounces of tea at 6¼ cts.? *Ans.* $9.75.
10. 896 skeins of silk at 5 cents? *Ans.* $44.8.
11. 4 dozen copy books at 10 cents? *Ans.* $4.80.
12. 364 bushels potatoes at 50 cts? *Ans.* $182.
13. 486 yds. cassimer at 10s.? *Ans.* £243.
14. 896 thirds of a £ at 6s. 8d.? *Ans.* £298 13s. 4d.
15. 764 fourths of a £ at 5s.? *Ans.* £191.
16. 654 fifths of a £ at 4s. each? *Ans.* £130 16s.
17. 726 sixths of a £ at 3s. 4d. each? *Ans.* £121.
18. 583 eighths of a £ at 2s. 6d.? *Ans.* £72 17s. 6d.

Case 3. When the price is the complement of a single part of the money unit, and the quantity a simple number; divide the quantity by the denominator of the part, as in case 2, and subtract the quotient from the quantity taken as a multiple of the money unit; the remainder will be the answer.

Examples for the Slate.

1. 5671 pairs at 66⅔ cts., and at 13s. 4d.

cts.		s. d.			
	5671.00		5671	0	0
66⅔, comp. ⅓	1890.33⅓	13 4, comp. ⅓	£1890	6	8
	Ans. $3780.66⅔		*Ans.* £3780	13	4

2. 375 pieces of wall paper at 75 cts.? *Ans.* $281.25.
3. 846 lbs. Imperial tea at 80 cts.? *Ans.* $676.80.
4. 739 lbs. Gunpowder tea at 83⅓ cts.? *Ans.* $615.83⅓.
5. 836 gallons sperm oil at 87½ cts.? *Ans.* $731.50.
6. 584 bushels of rye at 90 cts.? *Ans.* $525.60.
7. 468 yards vesting at 91⅔ cts.? *Ans.* $429.
8. 396 boxes of pipes at 93¾ cts.? *Ans.* $371.25.
9. 254 yds. oil cloth at 95 cts.? *Ans.* $241.3.
10. 654 merino sheep at 15s.? *Ans.* £490 10s.

1101	1102	1103	1104	1105	1106	1107	1108	1109	1110
$\frac{1}{1101}$	$\frac{1}{1102}$	$\frac{1}{1103}$	$\frac{1}{1104}$	$\frac{1}{1105}$	$\frac{1}{1106}$	$\frac{1}{1107}$	$\frac{1}{1108}$	$\frac{1}{1109}$	$\frac{1}{1110}$

11. 432 loads of wood at 16s.? *Ans.* £345 12s.
12. 842 bushels of rye at 93$\frac{3}{4}$ cts.? *Ans.* 789.37\frac{1}{2}$.
13. 956 bushels of wheat at 95 cts.? *Ans.* $908.2.
14. 243 bushels of rye at 87$\frac{1}{2}$ cts.? *Ans.* 212.62\frac{1}{2}$.
15. 432 bushels of corn at 75 cts.? *Ans.* $324.
16. 846 beaver hats at 17s. 6d.? *Ans.* £740 5s.
17. 958 yards of crape at 13s. 4d.? *Ans.* £638 13s. 4d.

Case 4. When the price of a unit is given, and the quantity is a single part of the unit; divide the price of the unit by the denominator of the part, for the answer.

Examples for the Slate.

1. 6 oz. troy at $13.96 cts. per lb.?

	$ cts.
	13.96
6 oz.=$\frac{1}{2}$	$6.98 *Ans.*

2. 4 oz. troy, or $\frac{1}{3}$ of 1 lb., at $13.96 per lb.? *Ans.* 4.65\frac{1}{3}$.
3. 3 oz. troy, or $\frac{1}{4}$ of 1 lb., at $13.96 per lb.? *Ans.* $3.49.
4. 2 oz. 8 dwts., or $\frac{1}{5}$ of 1 lb., at $13.96 per lb.? *Ans.* $2.792.
5. 2 oz. troy, or $\frac{1}{6}$ of 1 lb., at $13.96 per lb.? *Ans.* 2.32\frac{2}{3}$.
6. 1 oz. 10 dwts., or $\frac{1}{8}$ of 1 lb., at $13.96 per lb.? *Ans.* $1.745.
7. 1 oz. 4 dwts., or $\frac{1}{10}$ of 1 lb., at $13.96 per lb.? *Ans.* $1.396.
8. 1 oz., or $\frac{1}{12}$ of 1 lb., at $13.96 per lb.? *Ans.* 1.16\frac{1}{3}$.
9. 15 dwts., or $\frac{1}{16}$ of 1 lb., at $13.96 per lb.? *Ans.* $.87$\frac{1}{4}$.
10. 12 dwts., or $\frac{1}{20}$ of 1 lb., at $13.96 per lb.? *Ans.* $.698

1111	1112	1113	1114	1115	1116	1117	1118	1119	1120
$\frac{1}{1111}$	$\frac{1}{1112}$	$\frac{1}{1113}$	$\frac{1}{1114}$	$\frac{1}{1115}$	$\frac{1}{1116}$	$\frac{1}{1117}$	$\frac{1}{1118}$	$\frac{1}{1119}$	$\frac{1}{1120}$

11. The price of 1 ton of coal is \$5.75, how much will be the price of 10 cwt.?

\$ cts.
5.75

10 cwt=$\frac{1}{2}$ \$2.87$\frac{1}{2}$ *Ans.*

At \$5.75 per ton, what will—
12. 5 cwt., or $\frac{1}{4}$ of a ton, come to? *Ans.* \$1.43$\frac{3}{4}$.
13. 4 cwt., or $\frac{1}{5}$ of a ton, come to? *Ans.* \$1.15.
14. 2 cwt. 3 qrs. 12 lbs., $\frac{1}{7}$ T., come to? *Ans.* \$.82$\frac{1}{7}$.
15. 2 cwt. 2 qrs., or $\frac{1}{8}$ ton, come to? *Ans.* \$.71$\frac{7}{8}$.
16. 2 cwt., or $\frac{1}{10}$ of a ton, come to? *Ans.* \$.575.
17. 1 cwt. 1 qr. 20 lbs., $\frac{1}{14}$ T., come to? *Ans.* \$.41$\frac{1}{14}$.
18. 1 cwt. 1 qr., or $\frac{1}{16}$ ton, come to? *Ans.* \$.35$\frac{15}{16}$.
19. 1 cwt., or $\frac{1}{20}$ ton, come to? *Ans.* \$.287$\frac{1}{2}$.
20. 2 qrs. 24 lbs., or $\frac{1}{28}$ T., come to? *Ans.* \$.20$\frac{15}{28}$.
21. 20 lbs., or $\frac{1}{112}$ ton, come to? *Ans.* \$.05$\frac{15}{112}$.
22. The price of 1 cwt. of butter is \$28, how much should be the price of 56 lbs.?

\$ cts.
28.00

56 lbs.=$\frac{1}{2}$ \$14.00 *Ans.*

23. 2 qrs., or $\frac{1}{2}$ cwt., at \$28 per cwt.? *Ans.* \$14.
24. 1 qr., or $\frac{1}{4}$ cwt., at \$28 per cwt.? *Ans.* \$7.
25. 16 lbs., or $\frac{1}{7}$ cwt., at \$28 per cwt.? *Ans.* \$4.
26. 14 lbs., or $\frac{1}{8}$ cwt., at \$28 per cwt.? *Ans.* \$3.50.
27. 8 lbs., or $\frac{1}{14}$ cwt., at \$28 per cwt.? *Ans.* \$2.
28. 7 lbs., or $\frac{1}{16}$ cwt., at \$28 per cwt.? *Ans.* \$1.75
29. One pound of quinine sells at \$32; how much then is 6 oz. worth?

\$ cts.
32.00

6 oz.=$\frac{1}{2}$ \$16.00 *Ans.*

30. 4 oz. of quinine, at \$32 per lb.? *Ans.* \$10.66$\frac{2}{3}$.
31. 3 oz. of quinine, at \$32 per lb.? *Ans.* \$8.

1121	1122	1123	1124	1125	1126	1127	1128	1129	1130
$\frac{1}{1121}$	$\frac{1}{1122}$	$\frac{1}{1123}$	$\frac{1}{1124}$	$\frac{1}{1125}$	$\frac{1}{1126}$	$\frac{1}{1127}$	$\frac{1}{1128}$	$\frac{1}{1129}$	$\frac{1}{1130}$

32. 2 oz. of quinine, at $32 per lb.? *Ans.* $5.33⅓.
33. 1 oz. 4 drs. of quinine at $32 per lb.? *Ans.* $4.
34. 1 oz. of quinine, at $32 per lb.? *Ans.* $2.66⅔.
35. 6 drs. of quinine, at $32 per lb.? *Ans.* $2.
36. The cost of making 1 mile of railroad was $12680, how much then should 4 furlongs cost?

$12680.00

4 fur.=½ $6340.00 *Ans.*

37. 2 fur. 26⅔ po., or ⅓ m., at $12680? *Ans.* $4226⅔.
38. 2 fur., or ¼ m., at $12680? *Ans.* $3170.
39. 1 fur. 24 po., or ⅕ m., at $12680? *Ans.* $2536.
40. 1 fur., or ⅛ m., at $12680? *Ans.* $1585.
41. 32 po., or $\frac{1}{10}$ m., at $12680? *Ans.* $1268.
42. 20 po., or $\frac{1}{16}$ m., at $12680? *Ans.* $792.5.
43. 16 po., or $\frac{1}{20}$ m., at $12680? *Ans.* $634.
44. 13 po. 5⅓ ft., $\frac{1}{24}$ m., at $12680? *Ans.* $528.33⅓.
45. 10 po., or $\frac{1}{32}$ m., at $12680? *Ans.* $396.25.
46. 1 yard of broadcloth cost $2.88, how much then is the cost of 2 quarters, or 1 ft. 6 in.?

$2.88

2 qrs.=½ $1.44 *Ans.*

47. 1 ft., or ⅓ yd., at $2.88 per yd.? *Ans.* $.96.
48. 1 qr., 9 in., or ¼ yd., at $2.88 per yd.? *Ans.* $.72.
49. 6 in., or ⅙ yd., at $2.88 per yd.? *Ans.* $.48.
50. 2 na., 4½ in., ⅛ yd., at $2.88 per yd.? *Ans.* $.36.
51. 4 in., or ⅑ yd., at $2.88 per yd.? *Ans.* $.32.
52. 3 in., or $\frac{1}{12}$ yd., at $2.88 per yd. *Ans.* $.24.
53. 2¼ in., 1 na., $\frac{1}{16}$ yd., at $2.88 per yd.? *Ans.* $.18.
54. 1 square yard of ground in a populous city being valued at as many silver dollars as would cover it, and $1728 was agreed upon, how much was the value of 4 sq. ft. 72 sq. in. at that rate?

$1728.00

sq. ft. sq. in.
4 72=½ yd. $864.00 *Ans.*

1131	1132	1133	1134	1135	1136	1137	1138	1139	1140
$\frac{1}{1131}$	$\frac{1}{1132}$	$\frac{1}{1133}$	$\frac{1}{1134}$	$\frac{1}{1135}$	$\frac{1}{1136}$	$\frac{1}{1137}$	$\frac{1}{1138}$	$\frac{1}{1139}$	$\frac{1}{1140}$

55. 3 sq. ft., or $\frac{1}{3}$ of a sq. yd., at $1728? *Ans.* $576.
56. 2 sq. ft. 36 in., $\frac{1}{4}$ sq. yd., at $1728? *Ans.* $432.
57. 1 sq. ft. 72 in., $\frac{1}{6}$ sq. yd., at $1728? *Ans.* $2.88.
58. 1 sq. ft. 18 in., $\frac{1}{8}$ sq. yd., at $1728? *Ans.* $216.
59. 1 sq. ft., or $\frac{1}{9}$ sq. yd. at $1728? *Ans.* $192.
60. 108 sq. in., or $\frac{1}{12}$ sq. yd., at $1728? *Ans.* $144.
61. 81 sq. in., or $\frac{1}{16}$ sq. yd., at $1728? *Ans.* $108.
62. 72 sq. in., or $\frac{1}{18}$ sq. yd., at $1728? *Ans.* $96.
63. 36 sq. in., or $\frac{1}{36}$ sq. yd., at $1728? *Ans.* $48.

64. One acre of public land is sold for the use of the U. S. at $1.25; how much, at that rate, are 2 roods worth?

$ cts.
1.25

2 roods $=\frac{1}{2}$ A. 0.62\frac{1}{2}$ *Ans.*

At $1.25 per acre, what will—

65. 53$\frac{1}{3}$ sq. po., or $\frac{1}{3}$ A., come to? *Ans.* $.41$\frac{2}{3}$.
66. 40 sq. po., or $\frac{1}{4}$ A., come to? *Ans.* $.31$\frac{1}{4}$.
67. 20 sq. po., or $\frac{1}{8}$ A., come to? *Ans.* $.15$\frac{5}{8}$.
68. 16 sq. po., or $\frac{1}{10}$ A., come to? *Ans.* $.12$\frac{1}{2}$.
69. 7$\frac{3}{11}$ sq. po., or $\frac{1}{22}$ A., come to? *Ans.* $.05$\frac{15}{22}$.
70. 4 po., or $\frac{1}{40}$ A., come to? *Ans.* $.03$\frac{1}{8}$.
71. 2 po., or $\frac{1}{80}$ A., come to? *Ans.* $.01$\frac{9}{16}$.

72. One gallon of camphene sells for 64 cents, how much, then, should 2 quarts bring?

cts.
.64

2 qts.$=\frac{1}{2}$ gal. $.32 cents. *Ans.*

73. One qt.? one pt.? 2 gls.? at 64 cts. per gal.? *Ans.* 16 cts., 8 cts., 4 cts.

74. One bushel of salt cost 64 cts., how much is $\frac{1}{2}$ bu.? 2 pe.? 1 pe.? 4 qts.? 2 qts.?

75. At $1095 per year, how much is that for 6 months? for $\frac{1}{3}$ of a leap year? for 13 weeks? for $\frac{1}{5}$ of a common year? and for $\frac{1}{6}$ of 12 calendar months?

76. One cord of hickory wood is worth, in some places, $6, how much, then, should 64 cubic feet sell for?

1141	1142	1143	1144	1145	1146	1147	1148	1149	1150
$\frac{1}{1141}$	$\frac{1}{1142}$	$\frac{1}{1143}$	$\frac{1}{1144}$	$\frac{1}{1145}$	$\frac{1}{1146}$	$\frac{1}{1147}$	$\frac{1}{1148}$	$\frac{1}{1149}$	$\frac{1}{1150}$

how much is 32 cu. ft. worth? how much is 16 cu. ft. worth? also 8? 4? 2?

77. The digging of 1 cubic yard of rock at the bottom of a well cost $81, then how much will be the cost of 13½ cubic feet? of 9 cu. ft.? of 3 cu. ft.? of 1 cu. ft.?

Case 5. When the price of a unit is given, and the quantity is the complement of a single part of that unit, subtract the price of the part from the price of the unit, the remainder will be the price of the complement.

Examples for the Slate.

1. The price of 1 lb. of coined silver is $13.96 nearly, then how much will be the value of 8 oz. troy?

	$ cts.
	At 13.96 per lb.
8 oz.=comp. of ⅓	4.65⅓
Ans.	$9.30⅔

2. 9 oz. troy, or ¾ lb., at $13.96? *Ans.* $10.47.
3. 9 oz. 12 dwts., or ⅘ lb., at $13.96? *Ans.* $11.168.
4. 10 oz., or ⅚ lb., at $13.96? *Ans.* $11.63⅓.
5. 10 oz. 10 dwts., or ⅞ lb., at $13.96? *Ans.* $12.215.
6. 10 oz. 16 dwts., or $\frac{9}{10}$ lb., at $13.96? *Ans.* $12.564.
7. 11 oz., or $\frac{11}{12}$ lb., at $13.96? *Ans.* $12.79⅔.
8. 11 oz. 5 dwts., or $\frac{15}{16}$ lb., at $13.96? *Ans.* $13.08¾.
9. 11 oz. 8 dwts., or $\frac{19}{20}$ lb., at $13.96? *Ans.* $13.262.
10. If 1 ton of coal cost $5.75, how much should be the cost of 15 cwt.?

cwt.	$ cts.
	At 5.75 per ton.
15=comp. of ¼	1.43¾
	$4.31¼ *Ans.*

Note. In the following examples, *c* stands for complement; as, *c.* ⅛, complement of ⅛, viz., of 1 ton, &c.

11. 16 cwt. of coal (*c.*⅕) at $5.75? *Ans.* $4.60.

1151	1152	1153	1154	1155	1156	1157	1158	1159	1160
$\frac{1}{1151}$	$\frac{1}{1152}$	$\frac{1}{1153}$	$\frac{1}{1154}$	$\frac{1}{1155}$	$\frac{1}{1156}$	$\frac{1}{1157}$	$\frac{1}{1158}$	$\frac{1}{1159}$	$\frac{1}{1160}$

12. 17 cwt. 16 lbs. (*c.* $\frac{1}{7}$) at $5.75? *Ans.* 4.92\frac{6}{7}$.
13. 17 cwt. 2 qrs. (*c.* $\frac{1}{8}$) at $5.75? *Ans.* 5.03\frac{1}{8}$.
14. 18 cwt. (*c.* $\frac{1}{10}$) at $5.75? *Ans.*$5.17$\frac{1}{2}$.
15. 18 cwt. 2 qrs. 8 lbs. (*c.* $\frac{1}{14}$) at $5.75? *Ans.* 5.33\frac{13}{14}$.
16. 18 cwt. 3 qrs. (*c.* $\frac{1}{16}$) at $5.75? *Ans.* 5.39$\frac{1}{16}$.
17. 19 cwt. (*c.* $\frac{1}{20}$) at $5.75? *Ans.* 5.46$\frac{1}{4}$.
18. 19 cwt. 1 qr. 4 lbs. (*c.* $\frac{1}{28}$) at $5.75? *Ans.* 5.54\frac{13}{28}$.
19. 19 cwt. 3 qrs. 8 lbs. (*c.* $\frac{1}{112}$) at $5.75? *Ans.* 5.69\frac{97}{112}$.

20. 3 qrs. of 1 cwt. of glades butter, at 28 dollars per cwt.; how much is the amount?

	$28.00	price of 1 cwt.
3 qrs. comp. $\frac{1}{4}$	7.00	" " 1 qr.
Ans.	$21.00	" " 3 qrs.

21. 3 qrs. 12 lbs. (*c.* $\frac{1}{7}$) at $28 per cwt.? *Ans.* $24.
22. 3 qrs. 14 lbs. (*c.* $\frac{1}{8}$) at $28 per cwt.? *Ans.* $24.50.
23. 3 qrs. 20 lbs. (*c.* $\frac{1}{14}$) at $28 per cwt.? *Ans.* 26.
24. 3 qrs. 21 lbs. (*c.* $\frac{1}{16}$) at $28 per cwt.? *Ans.* $26.25.

25. 8 ounces of quinine (comp. $\frac{1}{3}$ lb.), at the rate of $32 per lb.; what does it amount to?

	$32.00
8 oz. comp. $\frac{1}{3}$ lb.	10.66$\frac{2}{3}$
	21.33\frac{1}{3}$ *Ans.*

26. 9 oz. of quinine (*c.* $\frac{1}{4}$) at $32 per lb.? *Ans* $24.
27. 10 oz. of quinine (*c.* $\frac{1}{6}$) at $32? *Ans.* 26.66\frac{2}{3}$.
28. 10 oz. 4 drs. of quinine (*c.* $\frac{1}{8}$) at $32? *Ans.* $28.
29. 11 oz. of quinine (*c.* $\frac{1}{12}$) at $32? *Ans.* 29.33\frac{1}{3}$.
30. 11 oz. 2 drs. of quinine (*c.* $\frac{1}{16}$) at $32? *Ans.* $30.

31. The construction of 1 mile of railroad cost $12680, then how much should be the expense of 5 furlongs, 13 poles, 5$\frac{1}{2}$ ft. of the same work?

fur.	po.	ft.		$ cts.
				12680.00
5	13	5$\frac{1}{2}$	comp. $\frac{1}{3}$	4226.66$\frac{2}{3}$
				8453.33\frac{1}{3}$ *Ans.*

1161	1162	1163	1164	1165	1166	1167	1168	1169	1170
$\frac{1}{1161}$	$\frac{1}{1162}$	$\frac{1}{1163}$	$\frac{1}{1164}$	$\frac{1}{1165}$	$\frac{1}{1166}$	$\frac{1}{1167}$	$\frac{1}{1168}$	$\frac{1}{1169}$	$\frac{1}{1170}$

32. 6 furlongs (*c.* $\frac{1}{4}$) at $12680 per m.? *Ans.* $9510.
33. 6 fur. 16 po. (*c.* $\frac{1}{5}$) at $12680? *Ans.* $10144.
34. 7 furlongs (*c.* $\frac{1}{8}$) at $12680? *Ans.* $11095.
35. 7 fur. 8 po. (*c.* $\frac{1}{10}$) at $12680? *Ans.* $11412.
36. 7 fur. 20 po. (*c.* $\frac{1}{16}$) at $12680? *Ans.* $11887.5.
37. 7 fur. 24 po. (*c.* $\frac{1}{20}$) at $12680? *Ans.* $12046.
38. 7 fur. 26 po. 11 ft. (*c.* $\frac{1}{24}$), at $12680 per m.?
39. 7 fur. 30 po. (*c.* $\frac{1}{32}$) at $12680 per m.?
40. If one yard of broadcloth cost $2.88, how much will 2 feet of the same stuff be worth?

	$ cts.
	2.88
2 feet, comp. $\frac{1}{3}$	.96
	$1.92

41. 3 qrs. of cloth (*c.* $\frac{1}{4}$) at $2.88 per yard? *Ans.* $2.16.
42. 30 in. (*c.* $\frac{1}{6}$) at $2.88 per yd.? *Ans.* $2.40.
43. 3 qrs. 2 na. (*c.* $\frac{1}{8}$) at 2.88 per yd.? *Ans.* $2.52.
44. 32 in. (*c.* $\frac{1}{9}$) at $2.88 per yd.? *Ans.* $2.56.
45. 33 in. (*c.* $\frac{1}{12}$) at $2.88 per yd.? *Ans.* $2.64.
46. 3 qrs. 3 na. (*c.* $\frac{1}{16}$), at $2.88 per yd.? *Ans.* $2.70.
47. If one square yard of ground, in a place of populous concourse, be worth $1728; how much will 6 sq. ft. of the same be rated at?

	$ cts
	1728.00
6 feet com. $\frac{1}{3}$	576.00
	$1152.00 *Ans.*

48. 6 sq. ft. 108 sq. in. (*c.* $\frac{1}{4}$) at $1728? *Ans.* $1296.
49. 7 sq. ft. 72 sq. in. (*c.* $\frac{1}{6}$) at $1728? *Ans.* $1440.
50. 7 sq. ft. 126 sq. in. (*c.* $\frac{1}{8}$) at $1728? *Ans.* $1512.
51. 8 sq. ft. (*c.* $\frac{1}{9}$) at $1728? *Ans.* $1536.
52. 8 sq. ft. 36 sq. in. (*c.* $\frac{1}{12}$) at $1728? *Ans.* $1584.
53. 8 sq. ft. 63 sq. in. (*c.* $\frac{1}{16}$) at $1728? *Ans.* $1620.
54. 8 sq. ft. 72 sq. in. (*c.* $\frac{1}{18}$) at $1728? *Ans.* $1632.
55. 8 sq. ft. 108 sq. in. (*c.* $\frac{1}{36}$) at $1728? *Ans.* $1680.

1171	1172	1173	1174	1175	1176	1177	1178	1179	1180
$\frac{1}{1171}$	$\frac{1}{1172}$	$\frac{1}{1173}$	$\frac{1}{1174}$	$\frac{1}{1175}$	$\frac{1}{1176}$	$\frac{1}{1177}$	$\frac{1}{1178}$	$\frac{1}{1179}$	$\frac{1}{1180}$

56. If one acre of United States land cost $1.25, how much should 3 roods of the same come to?

$ cts.
1.25
3 roods, comp. $\frac{1}{4}$ .31$\frac{1}{4}$

0.93\frac{3}{4}$ *Ans.*

57. 3 R. 20 po. (*c.* $\frac{1}{8}$) at $1.25 per A.? *Ans.* 1.09\frac{3}{8}$.
58. 3 R. 24 po. (*c.* $\frac{1}{10}$) at $1.25 per A.? *Ans.* 1.12\frac{1}{2}$.
59. 3 R. 30 po. (*c.* $\frac{1}{16}$) at $1.25 per A.? *Ans.* 1.17\frac{3}{16}$.
60. 3 R. 32 po. 22 yds. (*c.* $\frac{1}{22}$) at $1.25? *Ans.* 1.19\frac{7}{22}$.
61. 3 R. 36 po. (*c.* $\frac{1}{40}$) at $1.25? *Ans.* 1.21\frac{7}{8}$.
62. 3 R. 38 po. (*c.* $\frac{1}{80}$) at $1.25? *Ans.* 1.23\frac{7}{16}$.
63. If an estate yield exactly $1095 per annum; how much will that be in 8 mos. or $\frac{2}{3}$ of a common year?

$1095.00
8 mos, com. $\frac{1}{3}$ 365.00

$730.00 *Ans.*

Case 6. When the quantity is composed of plural parts of the unit of which the price is given; take out of it the greatest single part of the unit it contains; also the next greatest, and the next; and so on until the entire quantity is divided into parts of the unit, or of parts previously taken: then find by division, as in Case 4, the price of each part, and add the several prices into one sum for the answer.

Examples for the Slate.

1. What will 2 qrs. 14 lbs. come to, at $14.5 per cwt.?

	qrs.	lbs.	$14.5 per cwt.
Of 1 cwt.	2	0=$\frac{1}{2}$	7.25
Of 2 qrs.		14=$\frac{1}{4}$	1.81$\frac{1}{4}$
	2	14	9.06\frac{1}{4}$ *Ans.*

1181	1182	1183	1184	1185	1186	1187	1188	1189	1190
$\frac{1}{1181}$	$\frac{1}{1182}$	$\frac{1}{1183}$	$\frac{1}{1184}$	$\frac{1}{1185}$	$\frac{1}{1186}$	$\frac{1}{1187}$	$\frac{1}{1188}$	$\frac{1}{1189}$	$\frac{1}{1190}$

2. 2 qrs. 25 lbs. at $16.8 per cwt.? *Ans.* $12.15.
3. 13 oz. 8 drs. at $3. per lb.? *Ans.* $2.53125.
4. 3 R. 24 po. at $108 per acre? *Ans.* $97.2.
5. 37 gals. at $141.75 per hhd.? *Ans.* $83.25.
6. 12 cwt. 3 qrs. 24 lbs. at £56 per T.? *Ans.* £36 6s.

Case 7. When the quantity is composed of a multiple and plural parts of the unit of which the price is given; take such multiple and such parts of the price as the quantity is of the unit, and add them into one sum for the answer

Examples for the Slate.

1. 9 cwt. 2 qrs. 26 lbs. at $14.75 per cwt.?

	qrs. lbs.	
		$14.75 price of 1 cwt
		9
		132.75 9th multiple of the price.
Of 1 cwt.	2 0=$\frac{1}{2}$	7.375 price of 2 qrs.
Of 1 cwt.	16=$\frac{1}{7}$	2.107$\frac{1}{7}$ " of 16 lbs.
Of 16 lbs.	8=$\frac{1}{2}$	1.053$\frac{4}{7}$ " of 8 lbs.
Of 8 lbs.	2=$\frac{1}{4}$	.263$\frac{11}{28}$ " of 2 lbs.
		143.549\frac{3}{28}$ *Ans.*

2. 59 cwt. 1 qr. 14 lbs. at $7.50 per cwt.? *Ans.* 445.31\frac{1}{4}$.
3. 8 cwt. 1 qr. 16 lbs. at $12.25 per cwt.? *Ans.* 102.81\frac{1}{4}$.
4. 67 yds. 2 qrs. at $3.5 per yd.? *Ans.* $236.25.
5. 48 lbs. 6 oz. 6 dwts. at $18. per lb.? *Ans.* $873.45.
6. 42 lbs. 10 oz. 6 drs. 1 sc. 4 grs. at $12. per lb.? *Ans.* $514.8.
7. 149 E.E. 2 qrs. at £1 15s. per ell.? *Ans.* £261 9s.
8. 56 yds. 1 qr. at $6.25 per yd.? *Ans.* 351.56\frac{1}{4}$.
9. 48 E.Fl. 2 qrs. at 18$\frac{3}{4}$ cts. per ell? *Ans.* 9.12\frac{1}{2}$.
10. 598 A. 2 R. at $24.5 per A.? *Ans.* $14663.25.
11. 56 hhds. 42 gals. at 56\frac{1}{4}$ per hhd.? *Ans.* $3187.5.

1191	1192	1193	1194	1195	1196	1197	1198	1199	1200
$\frac{1}{1191}$	$\frac{1}{1192}$	$\frac{1}{1193}$	$\frac{1}{1194}$	$\frac{1}{1195}$	$\frac{1}{1196}$	$\frac{1}{1197}$	$\frac{1}{1198}$	$\frac{1}{1199}$	$\frac{1}{1200}$

12. 100 bu. 2 pe. at 93$\frac{3}{4}$ cts. per bu.? *Ans.* \$94.21$\frac{7}{8}$.

13. 24$\frac{3}{4}$ tons of coal, at 6 dollars 50 cents per ton; what is the amount? *Ans.* \$160.87$\frac{1}{2}$.

Note. But if the quantity composed of a multiple and plural parts, want only a single part to reach the next multiple; take that next multiple of the price, and subtract from it the single defective part; the remainder will be the answer.

Examples for the Slate.

1. 72 cwt. 3 qrs. 21 lbs. of feathers, at \$33.6 per cwt.?

```
cwt. qrs. lbs.          $33.60 price of 1 cwt.
 72   3   21+7 lbs.=        73
                        ------
                         10080
                        23520
                        ------
                        2452.80
Of 1 cwt. 7 lbs.=1/16=  — 2.10
                        ------
                      $2450.70 Ans.
```

2. 209 yds. 3 qrs. 2 na. at \$1.25 per yd.? *Ans.* \$262.34$\frac{3}{8}$.

3. 29 gals. 3 qts. 1 pt. at \$1.5 per gal.? *Ans.* \$44.81$\frac{1}{4}$.

4. 17 cwt. 3 qrs. 14 lbs. at \$15.5 per cwt.? *Ans.* \$277.06$\frac{1}{4}$.

5. 8 bu. 3 pe. 4 qts. at \$1.25 per bu.? *Ans.* \$11.09$\frac{3}{8}$.

6. 107 ft. 9 in. at \$2.28 per foot? *Ans.* \$245.67.

7. 34 yds. 3 qrs. 2 na. at 65 cts. per yd.? *Ans.* \$22.66$\frac{7}{8}$.

8. 43 A. 3 R. 20 po. at \$68 per A.? *Ans.* \$2983.5.

General Rule. The simplest rule, however, and that which includes every possible case, is that which reduces all the grades of units in the given quantity to that of the unit of which the price is given: the work is then a simple or decimal multiplication.

1201	1202	1203	1204	1205	1206	1207	1208	1209	1210
$\frac{1}{1201}$	$\frac{1}{1202}$	$\frac{1}{1203}$	$\frac{1}{1204}$	$\frac{1}{1205}$	$\frac{1}{1206}$	$\frac{1}{1207}$	$\frac{1}{1208}$	$\frac{1}{1209}$	$\frac{1}{1210}$

Examples for the Slate.

1. 5 years 3 months 25 days at $800 per annum? and 42 lbs. 10 oz. 6 drs. 1 sc. 4 grs. at $12 per lb.?

30)25.0 ds.	20) 4.0 grs.
12) 3.83333+mos.	3) 1.2 sc.
5.31944+yrs.	8) 6.4 drs.
800	12)10.8 oz.
	42.9 lbs.
$4255.55200 *Ans.*	at $12 per lb
	$514.8 *Ans.*

2. 3 qrs. 21 lbs. of pork at $5.5 per cwt.? *Ans.* $5.15625.

3. 12 oz. avoirdupois at $10.76 per lb.? *Ans.* $8.07.

4. $\frac{1}{3}$, $\frac{1}{7}$, and $\frac{1}{9}$ of a hhd. of wine are how many gallons? and how much the cost at $141.75 per hhd.? *Ans.* 37 gals. and $83.25.

5. 6 oz. 12 drs. avoirdupois at $12 per lb.? *Ans.* 5.06\frac{1}{4}$.

6. 22 cwt. 3 qrs. 27 lbs. of tanned leather at $20 per cwt; what is the amount? *Ans.* 459.82\frac{1}{7}$.

7. 144 cwt. 2 qrs. 21 lbs. of coffee at $9.3 per cwt.; how much does it amount to? *Ans.* $1345.59375.

8. What is the amount of 10 pieces of linen, each 22 yds. 2 qrs. 2 na., at 37$\frac{1}{2}$ cts. per yd.? *Ans.* 84.84\frac{3}{8}$.

RATES OF TARE.

Regulations respecting tares were enacted by Congress, and approved by the President, 2d March 1799. These are still in force at the several Custom-houses of the U. States.

Tare is the weight of the barrel, box, cloth, mat, or other envelope, in which the goods are packed.

Tare on sugar, not loaf, in hhds. and tierces, 12 per cent., or 12 lb.

1211	1212	1213	1214	1215	1216	1217	1218	1219	1220
$\frac{1}{1211}$	$\frac{1}{1212}$	$\frac{1}{1213}$	$\frac{1}{1214}$	$\frac{1}{1215}$	$\frac{1}{1216}$	$\frac{1}{1217}$	$\frac{1}{1218}$	$\frac{1}{1219}$	$\frac{1}{1220}$

in 100; in boxes 15 per cent.; in bags, or mats, 5 per cent.; in flour barrels 22 lbs. on each; on sugar candy, in boxes, 10 per cent.

Tare on coffee, in flour barrels, 20 lbs. each; in bags, or mats, 2 per cent. On pepper, in bags or mats, 2 per cent.; on pimento, 3 per cent. On cassia, in mats, 9 per cent.

Tare on cotton, in bales, 2 per cent.; in seroons, 6 per cent. On rice, in tierces and half tierces, 10 per cent. On copperas, in hhds., 10 per cent. On indigo, in seroons, 11 per cent.

On green teas 19 lbs. per chest; 12 lbs. per half chest. On souchong 22 lbs. per chest. On candles, in boxes, 8 per cent. On chocolate, in boxes, 10 per cent. On Glauber salts, and nails, in casks, 8 per cent. On soap, in boxes, 10 per cent. On shot, in casks, 3 per cent. On twine, in casks, 12 per cent.; in bales, 3 per cent.

On the same articles, in other packages, and upon all other articles subject to tare, the actual weight of the package is deducted. This practice is now prevailing among merchants.

There is a draught allowed at the Custom-houses upon articles subject to duty by weight, viz.: upon 1 cwt. 1 lb. is allowed; from 1 to 2 cwt., 2 lbs.; from 2 to 3 cwt., 3 lbs.; from 3 to 10 cwt., 4 lbs.; from 10 to 18 cwt., 7 lbs.; and upon 18 cwt. and above that, 9 lbs. are allowed.

Case 1. When equal tare is allowed on each package.

Rule. Multiply the tare on one package by the number of them, and subtract the product from the gross weight; the remainder will be the neat weight.

Examples for the Slate.

1. What is the neat weight of 20 barrels of sugar, the entire gross weight of which is 57 cwt. 1 qr. 22 lbs., tare on each barrel being 22 lbs.?

	cwt.	qrs.	lbs.	
Gross	57	1	22	
Tare	3	3	20	
Neat	53	2	2	*Ans.*

22 lbs. tare on each.
20

112)440(3 cwt.
336

28)104(3 qrs.
84

20 lbs.

2. What is the neat weight of 72 barrels of sugar, weighing each 2 cwt. 3 qrs. 14 lbs. gross; tare being allowed at 22 lbs. per barrel? *Ans.* 192 cwt. 3 qrs. 12 lbs.

1221	1222	1223	1224	1225	1226	1227	1228	1229	1230
$\frac{1}{1221}$	$\frac{1}{1222}$	$\frac{1}{1223}$	$\frac{1}{1224}$	$\frac{1}{1225}$	$\frac{1}{1226}$	$\frac{1}{1227}$	$\frac{1}{1228}$	$\frac{1}{1229}$	$\frac{1}{1230}$

3. What is the neat weight of 28 barrels of coffee weighing each 1 cwt. 2 qrs. 16 lbs. gross; tare on each 20 lbs.? *Ans.* 41 cwt.

4. How much is the neat weight of 32 chests of souchong tea, weighing each 80 lbs., at the regular tare of 22 lbs. on each; and how much is the cost, at $\frac{9}{16}$ of a dollar per lb.? *Ans.* 1856 lbs. weight, $1044 cost.

5. What is the neat weight of 12 chests of green tea, the gross being 68 lbs. each, and tare 19 lbs. each; and 12 half chests weighing 38 lbs. gross, tare on each 12 lbs.? *Ans.* 900 lbs.

6. How much is the value of 6 barrels of prunes, the whole gross weight being 16 cwt. 3 qrs. 21 lbs., tare per barrel, the same as on sugar in flour barrels; at the rate of 12½ cents per lb.? *Ans.* $220.62½.

The class will see, if they will consider, in the examples above, where the gross weights are equal on each package, as well as the tare, a shorter method than that prescribed in the rule.

Case 2. When the rate of tare is at so much per cent.

Rule. Reduce the entire quantity to lbs., and say as 100 is to the fixed rate of tare per cent. on the article, so is the quantity in lbs. to the tare on the whole.

Note. This reduction to lbs. is made necessary by the fact that the tare per cent. is given in lbs.

Examples for the Slate.

1. What is the neat weight of 10 hhds. of sugar, the sum of the whole gross weight being 114 cwt. 3 qrs. 14 lbs.; tare being allowed at 12 per cent.?

Reduce 114 cwt. 3 qrs. 14 lbs. to lbs.

11400=114×100 lbs. lbs. lbs.
1368=114×12 Then 1,00 : 12 : : 128,66 : x
98=3 qrs. 14 lbs. 12

12866 gross lbs. Tare 1543,92
1544 tare, nearly.

Ans. 11322 neat lbs.

1231	1232	1233	1234	1235	1236	1237	1238	1239	1240
$\frac{1}{1231}$	$\frac{1}{1232}$	$\frac{1}{1233}$	$\frac{1}{1234}$	$\frac{1}{1235}$	$\frac{1}{1236}$	$\frac{1}{1237}$	$\frac{1}{1238}$	$\frac{1}{1239}$	$\frac{1}{1240}$

2. At $7\frac{1}{4}$ cents per lb., how much is the value of 5 tierces of sugar, tare per cent. 12 lbs., the tierces being numbered and weighing as follows, viz.:

	cwt.	qrs.	lbs.	
No. 1.	5	3	18	gross.
2.	6	1	16	"
3.	7	2	8	"
4.	8	3	14	"
5.	9	2	18	"

Ans. 274.48\frac{1}{2}$.

3. The gross weights of 6 tierces of rice are as follows, viz.: 7 cwt. 2 qrs. 21 lbs., 8 cwt. 1 qr. 16 lbs., 9 cwt. 2 qrs. 14 lbs., 6 cwt. 1 qr. 8 lbs., 9 cwt. 3 qrs., and 8 cwt 2 qrs. 12 lbs.; and the regular tare is 10 per cent.: how much does the whole amount to, at $7\frac{1}{2}$ cents per lb. neat? *Ans.* 380.92\frac{1}{2}$.

4. Sold 175 cwt. 2 qrs. 24 lbs. of sugar, in hogsheads, allowing 12 per cent. tare; how much is the neat weight and value, at $6\frac{3}{4}$ cts. per lb. *Ans.* $1169.64.

5. Bought 39 cwt. 2 qrs. 14 lbs. of sugar, in Cuba boxes; tare being allowed at 15 per cent.; what is the neat weight, and the value at 10 cents per lb.? *Ans.* 3772.3 lbs. nt., $377.23.

6. How much is the neat weight and value of 3650 lbs. of coffee in mats at 2 per cent. tare, and at 9 cts. per lb. neat? *Ans.* 3577 lbs. nt., $321.93 value.

7. How much will 35 cwt. 3 qrs. 16 lbs. of rice neat, in tierces, the regular tare 10 per cent. being allowed? *Ans.* 3618 lbs.

8. What is the neat weight of 24 cwt. 2 qrs. 12 lbs. of indigo in seroons, tare at 11 per cent. being deducted? *Ans.* 2452.84 lbs.

9. Subtract the regular tare, 9 per cent., from 970 lbs. of cassia, in mats. *Ans.* 882.7 lbs. neat.

10. An importer entered at the custom house 200 hhds. of Porto Rico sugar, weighing gross 1746 cwt.; upon which he was allowed 9 lbs. for every draught of 18 cwt., and 12 per cent. tare: how much is the neat weight? *Ans.* 171307.52 lbs.

1241	1242	1243	1244	1245	1246	1247	1248	1249	1250
$\frac{1}{1241}$	$\frac{1}{1242}$	$\frac{1}{1243}$	$\frac{1}{1244}$	$\frac{1}{1245}$	$\frac{1}{1246}$	$\frac{1}{1247}$	$\frac{1}{1248}$	$\frac{1}{1249}$	$\frac{1}{1250}$

INTEREST, OR RATES PER CENT.

Simple Interest for 1 year, Commission, Insurance, Brokerage, and all Premiums at so much per cent., without reference to time, are subject to one general

RULE. Multiply the given sum by the given rate, and divide the product by 100.

In the calculations of Simple Interest the following terms are used, and the initial letters of them are helpful; namely:

p., the principal or sum lent.
C., the interest of 100 for 1 year, or rate per cent.
u., the yearly interest of 1, or rate per unum.
t., the time or number of years, months, days.
a., the amount of principal and interest.

The interest of 1 is derived from the rate per cent., by annexing 0s, and dividing by 100, see pages 44, 45, 46; and it is plain that having the interest of a unit, the interest of 2, 3, 4, 50, or any other number, may be found by multiplying it.

A Table of Rates per Centum and per Unum.

C.	*u.*	*C.*	*u.*	*C.*	*u.*	*C.*	*u.*
$\frac{1}{4}$	.0025	$2\frac{1}{4}$	.0225	$4\frac{1}{4}$	.0425	$6\frac{1}{4}$	.0625
$\frac{1}{2}$	.005	$2\frac{1}{2}$	.025	$4\frac{1}{2}$	.045	$6\frac{1}{2}$	.065
$\frac{3}{4}$	.0075	$2\frac{3}{4}$	.0275	$4\frac{3}{4}$	.0475	$6\frac{3}{4}$	.0675
1	.01	3	.03	5	.05	7	.07
$1\frac{1}{4}$	.0125	$3\frac{1}{4}$	.0325	$5\frac{1}{4}$	.0525	$7\frac{1}{4}$	.0725
$1\frac{1}{2}$	.015	$3\frac{1}{2}$	.035	$5\frac{1}{2}$	.055	$7\frac{1}{2}$	.075
$1\frac{3}{4}$	.0175	$3\frac{3}{4}$	.0375	$5\frac{3}{4}$	.0575	$7\frac{3}{4}$	.0775
2	.02	4	.04	6	.06	8	.08

In order to press the importance of this first step upon the class, if the reductions of quarters and halves to decimals, and the divisions by 100, have been too slightly regarded hitherto, let the numbers in the columns

1251	1252	1253	1254	1255	1256	1257	1258	1259	1260
$\frac{1}{1251}$	$\frac{1}{1252}$	$\frac{1}{1253}$	$\frac{1}{1254}$	$\frac{1}{1255}$	$\frac{1}{1256}$	$\frac{1}{1257}$	$\frac{1}{1258}$	$\frac{1}{1259}$	$\frac{1}{1260}$

above, marked *C.*, be reduced to the form in the columns marked *u.*, by the rules already given.

Since letters denote words, and may be conveniently used in *formulæ*, presenting at one view the rule and mode of operation, it is proper to identify them with the things which they represent.

The class will therefore answer the following questions, viz.:

What term or word does *p.* represent?
What term does *C.* represent?
What term does *u.* represent?
What term does *t.* represent?
What term does *a.* represent?
What operation does $C.\times t.$ indicate?
What operation does $u.\times t.$ indicate?
What operations do $\overline{C.\times t.\times p.}\div 100$ indicate?
What operation does $u.\times t.\times p.$ indicate?
Why is $\overline{C.\times t.\times p.}\div 100 = u.\times t.\times p.$?

Because $u. = C.\div 100$; that is, the quotient of the rate per cent. divided by 100 is equal to the rate per unit, or the interest of 1 for 1 year. This is the answer to the last question only.

Case 1. When the principal, *p.*, rate per cent. *C.*, and time, *t.*, are given to find the interest, and consequently the amount, *a.*; because $p.+\text{int.}=a.$

RULE. Since *C.* represents the interest of $100, and *u.* represents the interest of $1 for a year; therefore $C.\times t.=$ the interest of 100, and $u.\times t.=$the interest of 1 for the whole time.

And $100 : C.\times t.$, or $1 : u.\times t. :: p. :$ whole interest.

Wherefore, to find the 4th proportional, by rule page 96, we shall have $\overline{C.\times t.\times p.}\div 100$, or $u.\times t.\times p.=$whole interest: then, also, $p.+$whole interest$=a.$

Note. The time, according to this rule, is supposed to be in years; to which, if necessary, the days and months may be reduced, by the rule on page 89. The money unit may be $1, £1, 1 livre, or any denomination whatever.

The letters annexed to the terms in the examples below will explain the rule.

Examples for the Slate.

1. What is the amount of $32.5 for 2 years, 4 months, 24 days, at 6 per cent. per annum?

1261	1262	1263	1264	1265	1266	1267	1268	1269	1270
$\frac{1}{1261}$	$\frac{1}{1262}$	$\frac{1}{1263}$	$\frac{1}{1264}$	$\frac{1}{1265}$	$\frac{1}{1266}$	$\frac{1}{1267}$	$\frac{1}{1268}$	$\frac{1}{1269}$	$\frac{1}{1270}$

dols.
32.5 *p.*
2.4 *t.*

1300
650

78.00
6÷100=.06 *u.*

$4.68 whole interest.
32.5 +*p.*

Ans. $37.18 *a.*

30)24.0 ds.
12) 4.8 mos.
2.4 yrs.
See rule, page 89.

C.×*t.* *p.*
100 : 6×2.4 : : 32.5 : w. int.

or thus,
u×*t.* *p.*
1 : .06×2.4 : : 32.5 : w. int.
See rule, page 96.

2. What is the interest of $450 for two years, at 5 per cent. per annum? *Ans.* $45.

3. How much is the simple interest of $476 for 4 years, at 5 per cent. per annum? *Ans.* $95.2.

4. What is the interest of 72£ 12s. 6d. for 8 years, at 4½ per cent. per annum?

4½ *C.*
8 *t.* 12) 6.0 d.
20)12.5 s.
100 : 36 : : 72.625£ : w. int.
36

435750
217875

£26.14.500
20 · ·

s. 2.900
12 · ·

d. 10.8

In the product marked £, 4 occupies the units' rank; but division by 100 removes it to 6 in the rank of 100s. The fraction .145£×20 is reduced to shillings, making 2s. and .9 parts of a shilling. .9s.×12 makes 10d. and .8d. See "Reading of Decimals," page 37.

5. What is the simple interest of $384, at 6 per cent. per annum, for 4 years? *Ans.* $92.16.

1271	1272	1273	1274	1275	1276	1277	1278	1279	1280
$\frac{1}{1271}$	$\frac{1}{1272}$	$\frac{1}{1273}$	$\frac{1}{1274}$	$\frac{1}{1275}$	$\frac{1}{1276}$	$\frac{1}{1277}$	$\frac{1}{1278}$	$\frac{1}{1279}$	$\frac{1}{1280}$

6. How much is the simple interest of $432, for 5 years, at 6 per cent. per annum? *Ans.* $129.6.

7. What is the simple interest of $576, for 2 years, at 5 per cent. per annum? *Ans.* 57.6.

8. How much will $560 amount to in 4 years, at 6 per cent. per annum? *Ans.* $13.44.

9. What will a bond for $350 amount to, if allowed to remain at interest 3 yrs. 4 mos. 24 days, at $6.5 per cent. per annum? *Ans.* $427.35.

10. If $880 remain at interest 15 months, at 3.5 per cent. per annum, what sum will it have gained?
Ans. $38.5.

11. A borrowed $180, which he retained for 9 months, at the end of which time he paid the principal, and interest at 6 per cent. per annum; what was the amount?
Ans. $188.1.

12. What will be the amount of a bond, at the end of 11 yrs. 9 mos. 21 ds., the principal being $550, and the rate of interest 5 per cent. per annum? *Ans.* 874.729\frac{1}{6}$.

13. What interest may be demanded for the use of $184 for 3 mos. 18 days, where the interest is 7 per cent. per annum? *Ans.* $3.864.

14. How much will $50 gain in 8 months, at 5 per cent. per annum? *Ans.* 1.66\frac{2}{3}$.

15. How much will $100 gain in 73 days, where the interest is 8 per cent. per annum?

16. What will $100 gain in 292 days, where the interest is 8 per cent. per annum?

Note. When partial payments are made upon a note or bond, each payment and its date are indorsed upon the paper; the interest is computed down to that date; the sum paid is deducted from the amount of principal and interest, and the part unpaid continues to bear interest as before.

In this case the interest is computed for days; and the ratio is composed of two ratios, one of $100, £s, &c., compared with the principal; the other of 365 compared with the days for which computation is to be made: the rate per cent. per annum is the unmatched antecedent.

1281	1282	1283	1284	1285	1286	1287	1288	1289	1290
$\frac{1}{1281}$	$\frac{1}{1282}$	$\frac{1}{1283}$	$\frac{1}{1284}$	$\frac{1}{1285}$	$\frac{1}{1286}$	$\frac{1}{1287}$	$\frac{1}{1288}$	$\frac{1}{1289}$	$\frac{1}{1290}$

17. H held N's note for $200, dated 10th May, 1845; there were indorsed upon it as follows, viz.: "July 22, 1845, paid $56." "Dec. 15, 1845, paid $112." What amount remained unpaid at the last date, interest computed at 6 per cent. per an.?

The time for which the first interest is to be computed is 73 days, and the principal $200. Then we have

$100 : 200$: : 1 : 2
ds. 365 : 73 ds. : : 5 : 1
— — $ $
5 : 2 : : 6 : 2.4 First Int.

Therefore $200+2.4—56=202.4—56=146.4 dols. remained unpaid on the 22d July, 1845; from which date until Dec. 15th, 1845, there are 146 days.

Hence, again, we have

$100 : 146.4$: : 1 : 1.464
ds. 365 : 146 ds. : : 5 : 2
— —— $ $
Then 146.4 5 : 2.928 : : 6 : 3.5136
+ 3.5136 6
——— ———
149.9136 5)17.568
112. ———
——— $3.5136 Second interest.
$37.9136 Balance unpaid, Dec. 15. 1845.

18. A held B's bond for $450, bearing interest at 6 per cent., and dated October 20th, 1846. Payments indorsed "Jan. 1, 1847, paid $145.4." "Oct. 19, 1847, paid $160.88." How much remained due at the date of the last payment? *Ans.* $164.

Examples in Commission, Insurance, Brokerage.

19. What is the commission on the sale of goods amounting to $1388 at 10 per cent.? *Ans.* $138.8.

20. What is the insurance on property exposed to great risk, amounting to $136000, at the rate of $33\frac{1}{3}$ per cent.?

21. A broker sold 12 hhds. containing 125 cwt. of

1291	1292	1293	1294	1295	1296	1297	1298	1299	1300
$\frac{1}{1291}$	$\frac{1}{1292}$	$\frac{1}{1293}$	$\frac{1}{1294}$	$\frac{1}{1295}$	$\frac{1}{1296}$	$\frac{1}{1297}$	$\frac{1}{1298}$	$\frac{1}{1299}$	$\frac{1}{1300}$

sugar, for $8.4 per cwt., and charges $1\frac{3}{4}$ per cent. for his fees ; how much is he entitled to retain of the proceeds?

22. What should be paid for insuring $250.4 at 3.75 per cent.? *Ans.* $9.39.

23. How much is the brokerage of $540, at $37\frac{1}{2}$ cents per cent.? *Ans.* $$2.02\frac{1}{2}$.

24. The commission on 3728.4 dollars, is $4\frac{3}{4}$ per cent.; what is the amount? *Ans.* $177.099.

25. A ship and cargo bound for Russia is insured at 3 per cent. for $46380 ; how much is the amount?
Ans. $1391.4.

Rule. When the rate is 6 per cent. per annum, that is, when 100 brings 6 in 12 months ; 100 will bear 5 in 10 months, 4 in 8 months, 3 in 6 months, 2 in 4 months, and so on, viz.: the interest of 100 will always be equal to half the time reduced to months. But 100 is to its interest as any other sum is to its interest, that is,

$$100 : \tfrac{1}{2}\text{ m} : : p. : \text{whole interest.}$$

Examples for the Slate.

1. How much is the interest of $250 for 14 months, at 6 per cent. per annum?

$$100 : \tfrac{1}{2}\text{ of } 14 : : 250 : \text{whole interest.}$$
$$\tfrac{1}{2}\text{ of } 14 = \quad 7$$

Ans. $17.50

Note. If the rate were 5 per cent. instead of 6, find the interest for 6 per cent., and because 5 is the complement of one-sixth of 6, deduct one-sixth from the interest found for 6 per cent. and so, for all the multiples, parts, and complements of 6.

2. What is the interest of $7342 for $16\frac{1}{2}$ months at 6 per cent. per annum? *Ans.* $605.715.

3. What is the interest of $987.78 for 18 months and 18 days at 6 per cent. per annum? *Ans.* $91.86354.

4. How much will $156 amount to in 2 years 2 months 24 days, at 6 per cent. per annum? *Ans.* $176.904.

5. Let £25 10s. be at interest 3 years 3 months 12 days; what will be the amount at 5 per cent. per annum?
Ans. £29 13s. 8.7d.

1301	1302	1303	1304	1305	1306	1307	1308	1309	1310
$\frac{1}{1301}$	$\frac{1}{1302}$	$\frac{1}{1303}$	$\frac{1}{1304}$	$\frac{1}{1305}$	$\frac{1}{1306}$	$\frac{1}{1307}$	$\frac{1}{1308}$	$\frac{1}{1309}$	$\frac{1}{1310}$

6. What is the interest of $327.825, for 5 yrs. 5 months 27 days, at $4\frac{1}{2}$ per cent. per annum? *Ans.* $81.01.+

7. How much will £42 18s. 6d. amount to in 5 months 12 days, at 3 per cent. per annum? *Ans.* £43 10s. 1d.

8. What is the interest of $240, for 1 year 3 months 21 days, at 4 per cent. per annum? *Ans.* $12.56.

9. Required the interest of $1090, at 6 per cent. for 17 months? *Ans.* $92.75.

10. How much is the interest of $300, for 64 months, at $3\frac{3}{4}$ per cent. per annum? *Ans.* $60.

11. What is the interest of £416 12s. 6d. for 10 months, at 6 per cent. per annum? *Ans.* £20 16s. $7\frac{1}{2}$d.

Case 2. When the amount, $a.$, time, $t.$, and principal, $p.$, are given, to find the rate, $u.$ or $C.$

Rule. $t \times p : a-p :: 1 : u$, or as $100 : C$. Or, the product of the time and principal is to the difference of the amount and principal, as 1 is to the interest of 1, or 100 to the interest of 100 for 1 year.

The reason of this rule is obvious; for $t \times p$ is such a principal as will bear the whole interest $a-p$ in 1 year, as 1 bears the interest of 1, or 100 the interest of 100.

Examples for the Slate.

1. At what rate per cent. per annum will $200 amount to $260 in 5 years?

$p.$ 200$ 260 $a.$
$t.$ 5 yrs. 200 $p.$

$$1000 : 60 :: 100 : 6 = C.$$
$$\text{or } 1000 : 60 :: 1 : .06 = u.$$

2. At what rate of interest has $500 increased to $545 in 2 years? *Ans.* $4\frac{1}{2}$ per cent.

3. Suppose 440£ should increase to 481£ 5s. in $2\frac{1}{2}$ years, at what rate per cent. is the interest computed? *Ans.* $3\frac{3}{4}$ per cent.

4. At what rate per cent. will $300 amount to $396 in 8 years? *Ans.* 4 per cent.

1311	1312	1313	1314	1315	1316	1317	1318	1319	1320
$\frac{1}{1311}$	$\frac{1}{1312}$	$\frac{1}{1313}$	$\frac{1}{1314}$	$\frac{1}{1315}$	$\frac{1}{1316}$	$\frac{1}{1317}$	$\frac{1}{1318}$	$\frac{1}{1319}$	$\frac{1}{1320}$

Case 3. When the amount, *a.*, principal, *p.*, and rate, *u.* or *C.*, are given, to find the time *t.*

RULE. $p \times u : a - p :: 1 : t$. Wherefore, the whole interest $(a - p)$ divided by the interest of the principal $(p \times u)$ for 1 year will equal the time.

This rule says simply this, that the year's-interest is to the whole interest as 1 year is to the time.

In class: What does $p \times u$ represent? What does $a - p$ represent? How is a 4th proportional found? See p. 96.

Examples for the Slate.

1. In what time will $500 increase to $545, at $4\frac{1}{2}$ per cent. per annum? *Ans.* 2 years.

```
$500 p.
 .045 u.
 ------
 2500     $545 a
 2000      500 p.
 -----    ----
22.5,00 :  45  : : 1 yr. : 2 yrs.
```

2. In what time will $200 amount to $260, at 6 per cent. per annum? *Ans.* 5 years.

3. At the rate of $3\frac{3}{4}$ per cent., in what time will 440£ increase to 481£ 5s. *Ans.* 2.5 years.

4. In what time will $300 amount to $396, at 4 per cent. per annum? *Ans.* in 8 yrs.

5. In what time will $100 double itself at 6 per cent. per annum? also, at 12 per cent. per annum? and again at 3 per cent. per annum? *Ans.* at 6 per cent. in $16\frac{2}{3}$ yrs.

Case 4. When the amount, *a.*, time, *t.*, and rate, *u.* or *C.*, are given, to find the principal, *p.*

RULE. $1 + (u \times t) : a :: 1 : p$. That is, the amount of a unit is to the amount of the principal, as a unit is to the principal.

Here it is plain that *u.*, the interest of 1 for a year, multiplied by *t.*, the time, and the product added to 1, is the amount of a unit, in the same sense that *a.* is the amount of *p.*

This case of interest is usually called Discount; the amount is

1321	1322	1323	1324	1325	1326	1327	1328	1329	1330
$\frac{1}{1321}$	$\frac{1}{1322}$	$\frac{1}{1323}$	$\frac{1}{1324}$	$\frac{1}{1325}$	$\frac{1}{1326}$	$\frac{1}{1327}$	$\frac{1}{1328}$	$\frac{1}{1329}$	$\frac{1}{1330}$

some payment to be made in future, and the principal is the present worth of it.

The rule may also be varied as follows:

Rule. The amount of 100 at the given rate and time is to 100 as the sum to be discounted is to the present worth.

Note. The discount is found by subtracting the present worth from the amount; but more directly by the following

Rule. $100+C\times t : C\times t :: a :$ discount.

Examples for the Slate.

1. What principal at interest for 5 years will amount to $390? Or what is the present worth of $390 payable at the end of 5 years, discounting at 6 per cent. per annum.

$$\begin{array}{l} .06\ u. \\ \ \ 5\ t. \\ \hline \end{array}$$

$$1+.30 : \overset{a}{390} :: 1 : \overset{p.\,w.}{300}\ \textit{Ans.}$$

Or thus, $100+(6\times5) : 100 :: 390 : 300$.

2. What is the present worth of $272.5, payable in 2 years, discounting at 4½ per cent. per annum? *Ans.* $250.

3. What principal will increase itself to $481.25, in 2½ years, at 3¾ per cent. per annum? *Ans.* $440.

4. What should be paid in ready money for $240, due at the end of 8 years, allowing the purchaser 7½ per cent. per annum on his advance? *Ans.* $150.

5. What is the rebate or discount of $800, payable in 10 months, at 6 per cent. per annum? *Ans.* $38.096.

6. What is the difference between the interest and the discount upon $100 for 10 years, at 6 per cent. per annum? *Ans.* $22.5.

The discount is the interest of the present worth, or principal; and in the last example above it may be seen how considerably it differs from the interest of the amount; for the interest of the principal is the discount or rebate of the amount.

1331	1332	1333	1334	1335	1336	1337	1338	1339	1340
$\frac{1}{1331}$	$\frac{1}{1332}$	$\frac{1}{1333}$	$\frac{1}{1334}$	$\frac{1}{1335}$	$\frac{1}{1336}$	$\frac{1}{1337}$	$\frac{1}{1338}$	$\frac{1}{1339}$	$\frac{1}{1340}$

Bank Discount.

Notes discounted in banks are allowed to lie over three days, or until the third day, after their time expires, before payment is demanded, or protest entered; these are called days of grace, but they are always included in the computation of the discount, which is the simple interest of the sum for the time.

The banks reckon by days, 360 to the year, so that the borrower loses $\frac{1}{73}$ part of the time, besides paying interest upon that part of his note which the banks retain as discount.

RULE. Multiply the sum to be discounted by $\frac{1}{6}$ days, and divide by 1000; the quotient will be the discount.

Note. Division by 1000 requires only the removal of the units point to the rank of 1000s. The rule above depends upon an accidental relation existing between the terms employed in it; for as 100 in 360 days gives \$6, so (100×1-6th of 360)÷1000 gives 6 also. The rule is applicable to any number of days and sum.

Examples for the Slate.

1. How much is the discount on a note for **\$100**, payable in 63 days, including the days of grace?

$$\begin{array}{rr} & 100 \text{ dollars.} \\ \frac{1}{6} \text{ of } 63 \text{ days} = & 10\frac{1}{2} \\ \hline 1{,}000 & \$1{,}050. \end{array}$$

Ans. **\$1.05.**

2. What is the discount upon **\$520**, for 93 days, including the days of grace, at 6 per cent.? *Ans.* **\$8.06.**

3. What is the discount upon **\$960**, for 63 days, at 6 per cent.? *Ans.* 10.08.

4. What is the discount on **\$480**, for 25 days at **\$6** per cent.? *Ans.* **\$2.04.**

5. What is the discount upon **\$50**, for 59 days at 6 per cent.? *Ans.* **\$.491$\frac{2}{3}$.**

6. What is the discount upon **\$300**, for 78 days, at 6 per cent.? *Ans.* **\$3.9.**

7. What is the discount upon **\$500**, for 108 days, at 6 per cent.? *Ans.* **\$9.**

1341	1342	1343	1344	1345	1346	1347	1348	1349	1350
$\frac{1}{1341}$	$\frac{1}{1342}$	$\frac{1}{1343}$	$\frac{1}{1344}$	$\frac{1}{1345}$	$\frac{1}{1346}$	$\frac{1}{1347}$	$\frac{1}{1348}$	$\frac{1}{1349}$	$\frac{1}{1350}$

Class Exercise upon the four cases of Interest.

What are the peculiar terms or words used in Interest ?

Which are the given terms in the 1st case ?

Ans. p. principal ; *C.* rate per cent. ; *u.*, rate per unit ; *t.*, time.

100 is always understood to be given with *C.*, and 1 with *u.* ; that is, 100 and its rate, 1 and its rate, are implied in the letters.

Which are the given terms in the 2d case ?

Ans. a. t. p. Name the terms in full :

Which are the given terms in the 3d case ?

Ans. a. p. C. and *u.*

Which are the terms given in the 4th case ?

Ans. a. t. C. and *u.*

Which are the terms required in the 1st case ? in the 2d case ? in the 3d case ? in the 4th case ?

Construct each of the *formulæ* on slates ; read them, and recite them.

COMPOUND INTEREST.

When the simple interest, at the end of every term of payment, be it a year, ½ year, ¼ year, month, or day, is supposed to be joined to the principal, and both to bear interest for the following term, money is said to bear Compound Interest.

Let us here exhibit the common rule for computing Compound Interest, and reserve the subject for a more ample discussion upon the principles of involution and progressions.

Case 1. When the principal, rate and time are given to find the Compound Interest, or the amount.

Rule. Find the amount for the first year, or term of payment by Case 1, Simple Interest ; which consider as a new principal for the second year, or term of payment : find the amount for the second year in the same manner : and also for the third ; and so on through all the terms of payment : then from the last amount take the given principal ; the remainder will be the Compound Interest.

Note. If the rate per cent. be a single part of 100, take such part of each principal for the yearly interest.

1351	1352	1353	1354	1355	1356	1357	1358	1359	1360
$\frac{1}{1351}$	$\frac{1}{1352}$	$\frac{1}{1353}$	$\frac{1}{1354}$	$\frac{1}{1355}$	$\frac{1}{1356}$	$\frac{1}{1357}$	$\frac{1}{1358}$	$\frac{1}{1359}$	$\frac{1}{1360}$

Examples for the Slate.

1. What is the compound interest of $720 for 3 years, at 5 per cent. per annum?

$100 : 5 : : $720 1st. prin.

5	
36.00	1st. int.
720	
756	2d prin.
5	
37.80	2d int.
756	
793.8	3d prin.
5	
39.690	3d int.
793.8	
833.49	amount.
720.	1st prin.
$113.49 *Ans.*	

	$720	1st prin.
$\frac{5}{100}=\frac{1}{20}=$	36	1st int.
	756	2d prin.
$\frac{1}{20}=$	37.8	2d int.
	793.8	3d prin.
$\frac{1}{20}=$	39.69	3d int.
	833.49	amount.
	720.	1st prin.
	$113.49 *Ans.*	

2. What will £50 amount to, in 5 years, at 5 per cent. per annum, Compound Interest? *Ans.* £63 16s. 3½d.

3. What is the compound interest of $1000, for 4 years, at 4½ per cent. per annum? *Ans.* $192.5.

4. What is the compound interest of £370 for 6 years, at 4 per cent. per annum? *Ans.* £98 3s. 4¼ d.

5. What is the amount of £50 in 5 years, at 5 per cent. per annum, payable half yearly? *Ans.* £64 1d.

Note. In this ex. the rate is a single part of 100, viz., one-twentieth; its half is also a single part, viz., one-fortieth; therefore, if one-fortieth of the principal be added 10 times (for there are 10 half-yearly payments in 5 years), the last amount less the given principal will be the compound interest.

1361	1362	1363	1364	1365	1366	1367	1368	1369	1370
$\frac{1}{1361}$	$\frac{1}{1362}$	$\frac{1}{1363}$	$\frac{1}{1364}$	$\frac{1}{1365}$	$\frac{1}{1366}$	$\frac{1}{1367}$	$\frac{1}{1368}$	$\frac{1}{1369}$	$\frac{1}{1370}$

6. What is the compound interest of $629 for 7 years, at 6 per cent. per annum? *Ans.* $316.78.

7. How much will $1256 amount to in 8 years, at 6 per cent. per annum, compound interest? *Ans.* $2001.863+

EQUATION OF DATES AND PAYMENTS.

RULE 1. In the equation of dates, the sum of two bills of parcels of different dates *is to* the whole line of days between the dates *as* any one of the bills *is to* a section of the line measured from the opposite point; this measure will reach the day upon which the note should be dated.

The sum of the two bills, and the mean date found as above, may be compared with a third bill and its date; and this result with a fourth and its date; and so on to any number.

Examples for the Slate.

1. A bought goods of B, viz., on the 1st of May, of the value of $150, and on the 21st of June, of the value of $250; and wishing to close the account by A's note, the date of it is required?

The sum of the two bills is $400, and the line of days from May 1st to June 21st, inclusive, counts 52 days. Therefore,

400 : 52 : : 250 : 32½ counting from May,
and 400 : 52 : : 150 : 19½ counting from June.

In either case, the 2d of June is marked for the date.

One of the statements is sufficient; and the first is to be preferred: because it takes the days in their natural order.

2. B bought goods of C, namely, April 1st $300 worth, June 1st $400 worth; and having agreed to close the account by B's note, the date of it is required? *Ans.* May 6th.

1371	1372	1373	1374	1375	1376	1377	1378	1379	1380
$\frac{1}{1371}$	$\frac{1}{1372}$	$\frac{1}{1373}$	$\frac{1}{1374}$	$\frac{1}{1375}$	$\frac{1}{1376}$	$\frac{1}{1377}$	$\frac{1}{1378}$	$\frac{1}{1379}$	$\frac{1}{1380}$

3. C bought goods of D, August 10th $600 worth, and 20th October $800 worth; and closing with D's note, the date is required? *Ans.* Sept. 20th.

4. D sold to E on the 11th of June, a lot of dry goods worth $360, and on the 11th of September another lot worth $480; and closing the account by E's note, the date of it is required? *Ans.* August 16th.

5. E sold to F on the 20th of March a quantity of sugar worth $300, and on the 1st of May $200 worth of coffee; also, on the 4th of June 400 bushels of wheat, at $$1\frac{1}{8}$ per bushel; and closing with F's note, the date of it is required? *Ans.* May 4th.

In the equation of payments two cases are to be considered, viz.: 1. when the amount is given, and the mean time sought. 2. When the time of payment is fixed, and the interest, and of course the amount for the time is required. The two operations mutually prove each other.

RULE 1. To find the mean time: multiply each payment by its time; divide the sum of the products by the sum of the payments; the quotient will be the equated time.

RULE 2. To find the amount: find the present worth of each payment at legal discount, for its own time; add the present worths into one sum; and find how much the amount of this sum will be at the end of the fixed time.

Examples for the Slate.

1. M owes N $100, payable at 4 months, and $50, payable at 6 months; but preferring a single payment, the mean time is required?

$$\begin{array}{l} 100\times4=400 \\ 50\times6=300 \\ \overline{150}\quad)\overline{700}(4\tfrac{2}{3}\text{ mos. } \textit{Ans.} \\ \qquad\quad\;\; 600 \\ \qquad\quad\;\; \overline{100}\;\; \tfrac{100}{150}=\tfrac{10}{15}=\tfrac{2}{3}. \end{array}$$

1381	1382	1383	1384	1385	1386	1387	1388	1389	1390
$\frac{1}{1381}$	$\frac{1}{1382}$	$\frac{1}{1383}$	$\frac{1}{1384}$	$\frac{1}{1385}$	$\frac{1}{1386}$	$\frac{1}{1387}$	$\frac{1}{1388}$	$\frac{1}{1389}$	$\frac{1}{1390}$

2. M owes N $100, payable at 4 months, and $50, payable at 6 months; but they agree to include both payments in a note payable at the end of 4$\frac{2}{3}$ months; discount at 6 per cent., what sum should the note express?

Ans. $150.

Simp. Int., Case 4. $1+u\times t : 1 :: a : p.\ w.$ Thus,
in payment 1, $1+u\times t=1+.06\times\frac{1}{3}$ yr.$=1.02$, amount of 1.
in payment 2, $1+u\times t=1+.06\times\frac{1}{2}$ yr.$=1.03$, amount of 1.

Therefore, $100\div1.02=98.04$, *p. w.* of 100.
And $50\div1.03=48.54$, *p. w.* of 50.

The sum of both, is $146.58, *p. w.*, which, at 6 per cent. for 4$\frac{2}{3}$ mos., by Case 1, Simp. Int.,=150$, and is the sum to be expressed in the note.

3. B owes C $600, of which $200 are to be paid in 3 months, $150 in 4 months, and $250 in 6 months; how much is the equated time of these payments?

Ans. 4$\frac{1}{2}$ months.

4. B owes C $200, payable in 3 months, $150 payable in 4 months, and $250, payable in 6 mos.; but they agree to include the three payments in one note payable in 4$\frac{1}{2}$ months; how much should the note express, interest at 6 per cent.? *Ans.* $600.

Fixing the time of payment sooner or later than 4$\frac{1}{2}$ months would vary the amount.

5. C owes D $80, due in 4 months, $150 due in 6 months, $200 due in 8 months, and $220 due in 10 mos.; what is the proper time for paying the whole at once?

Ans. 7 mos. 21 ds.

6. D owes E a certain sum, of which $\frac{1}{2}$ is payable in 3 months, $\frac{1}{3}$ in 4 months, $\frac{1}{6}$ in 9 months; required the equated time for paying the whole debt?

7. E owes F $300 due in 4 months, and $100 due in 8 months; but agreeing to include both payments in a note, required the equated time? *Ans.* 5 months.

8. E owes F $300 due in 4 months, and $100 due in 8 months; but agreeing to include both payments in a note payable in 5 months, how much should the note express, interest at 6 per cent.? *Ans.* $400.

1391	1392	1393	1394	1395	1396	1397	1398	1399	1400
$\frac{1}{1391}$	$\frac{1}{1392}$	$\frac{1}{1393}$	$\frac{1}{1394}$	$\frac{1}{1395}$	$\frac{1}{1396}$	$\frac{1}{1397}$	$\frac{1}{1398}$	$\frac{1}{1399}$	$\frac{1}{1400}$

AMERICAN EXCHANGE.

In this branch of exchange we include the money of the British Provinces, Mexico, the West Indies, and South America generally; and cannot omit the old colonial currencies of the United States, which are still the colloquial money in many places, notwithstanding the superior excellence of the terms cent and dollar.

General Rule for the reduction of old Currencies.

Reduce the dollar to shillings, halves, thirds, or 12ths of a shilling, so that it be a multiple of the unit of the denomination to which it is reduced; and reduce the £ to the same denomination: make the former the numerator and the latter the denominator of a fraction.

This fraction will express the $\$$ in parts of the £; and if inverted, it will express the £ in parts of the $\$$.

Note. The given pence and shillings may be reduced to decimal parts of a £, which rule is always applicable.

1. In the States of New England, and also in Virginia and Kentucky, the $\$$ is called six shillings; that is, $\frac{6}{20}$ or $\frac{3}{10}$ £.

Hence $\$\times\frac{3}{10}=\pounds$; and $\pounds\times\frac{10}{3}=\$$.

2. In New York and North Carolina the $\$$ is called 8 shillings; that is, $\frac{8}{20}$ or $\frac{4}{10}$ £.

Hence, $\$\times.4=\pounds$; and $\pounds\times\frac{10}{4}=\$$.

3. In New Jersey, Pennsylvania, Delaware, and Maryland the $\$$ is called 7s. 6d.; that is, $7\frac{1}{2}$-20ths$=\frac{15}{40}=\frac{3}{8}\pounds$

Hence, $\$\times\frac{3}{8}=\pounds$; and $\pounds\times\frac{8}{3}=\$$.

Note. Since 7s. 6d.=90d.=100 cts.; therefore, d$+\frac{1}{9}$d.=cents; and cents—$\frac{1}{10}$ cts.=pence.

4. In South Carolina and Georgia, the $\$$ is called 4s. 8d.; that is, $4\frac{2}{3}$-20ths$=\frac{14}{60}=\frac{7}{30}$ £.

Hence $\$\times\frac{7}{30}=\pounds$; and $\pounds\times\frac{30}{7}=\$$.

5. In Newfoundland, New Brunswick, Nova Scotia, and Canada, the $\$$ is called 5s., or $\frac{5}{20}=\frac{1}{4}$ £.

Hence $\$\times\frac{1}{4}=\pounds$; and $\pounds\times4=\$$.

1401	1402	1403	1404	1405	1406	1407	1408	1409	1410
$\frac{1}{1401}$	$\frac{1}{1402}$	$\frac{1}{1403}$	$\frac{1}{1404}$	$\frac{1}{1405}$	$\frac{1}{1406}$	$\frac{1}{1407}$	$\frac{1}{1408}$	$\frac{1}{1409}$	$\frac{1}{1410}$

6. In Jamaica and Bermudas the $ is called 6s. 8d. or $6\frac{2}{3}$–20ths=$\frac{20}{60}$=$\frac{1}{3}$ £.
Hence $×$\frac{1}{3}$=£, and £×3=$.

7. In the French West Indies generally, accounts are kept in livres, sols and deniers; which are the same and also divided, as £ s. d. sterling.

The Spanish reckon the dollar, in some places for $8\frac{1}{4}$, in others for 9 livres. Hence the deniers and sols being reduced to decimal parts of a livre,

$×$8\frac{1}{4}$, or $×9=liv.; and liv.×$\frac{4}{33}$, or liv.×$\frac{1}{9}$=$.

Examples for the Slate.

1. How many dollars are equivalent to £556 18s. 4d. in currency of New England and Virginia?

To multiply by a fraction, the numerator is the multiplier and the denominator the divisor.

```
12)4.0 d.
20)18.33⅓ s.
   556.916⅔ £, worth 10/3$ each.
        10
   ---------
3)5569.16⅔
   ---------
 $1856.38⁸⁄₉ Ans.
```

2. What sum in the currency of New England will 1856.38\frac{8}{9}$ amount to?

Note. If $\frac{3}{10}$ had been set down as .3, without the denominator, as a decimal usually is, the division by 10 would have been unnecessary.

```
$1856.38⁸⁄₉, each 3/10 £
          3
   ---------
10)5569.16⅔
   ---------
  £556.916⅔
         20
   ---------
   s. 18.33⅓
         12
   ---------
    d. 4.00
```

Ans. £556 18s. 4d.

1411	1412	1413	1414	1415	1416	1417	1418	1419	1420
$\frac{1}{1411}$	$\frac{1}{1412}$	$\frac{1}{1413}$	$\frac{1}{1414}$	$\frac{1}{1415}$	$\frac{1}{1416}$	$\frac{1}{1417}$	$\frac{1}{1418}$	$\frac{1}{1419}$	$\frac{1}{1420}$

3. How much Federal money is equal to £135 12s. 6d., in Virginia currency? *Ans.* $\$452.08\frac{1}{3}$.

4. Reduce 1548 dollars to N. York Currency? *Ans.* £619 4s.

5. How many dollars are equivalent to £436 19s. 6d., currency of North Carolina? *Ans.* $1092.4375.

6. What sum in the currency of New York will $\$626.83\frac{1}{3}$ make? *Ans.* £250 14s. 8d.

7. How many dollars are equivalent to £125 12s. 6d., in the currency of the Middle States? *Ans.* $335.

8. In $400 how many £s, Pennsylvania currency? *Ans.* £150.

9. In 1440 Delaware pence how many cents? *Ans.* 1600 cents.

10. In 2400 cents how many pence of the currency of Maryland? *Ans.* 2160 pence.

11. In £384 15s. 8d., currency of Georgia, how many dollars? *Ans.* $1649.0714.

12. How much South Carolina money is equivalent to $2500? *Ans.* £583 6s. 8d.

13. How many dollars are equivalent to £44 5s. 10d., currency of Georgia and South Carolina? *Ans.* $189.821428.

14. How many dollars are equivalent to £72 12s. 6d., currency of Canada? *Ans.* $290.50.

15. Reduce $290.5 to the currency of Nova Scotia? *Ans.* £72 12s. 6d.

16. How many dollars are equivalent to £290 10s., in the currency of New Brunswick? *Ans.* $1162.

17. Reduce £49, Jamaica currency, to Federal money? *Ans.* $147.

18. Reduce $294.5 to the currency of Bermudas? *Ans.* £98 3s. 4d.

19. Reduce £112 15s., currency of Barbadoes, to dollars, the $ being 6s. 3d.? *Ans.* $360.80.

20. In 1804 dollars, how many £s in the currency of Barbadoes, the £ being $\frac{16}{5}\$$? *Ans.* £563 15s.

21. Change 9654 livres, of $8\frac{1}{4}$ livres to the $, into Federal money? *Ans.* $\$1170.18\frac{2}{11}$.

1421	1422	1423	1424	1425	1426	1427	1428	1429	1430
$\frac{1}{1421}$	$\frac{1}{1422}$	$\frac{1}{1423}$	$\frac{1}{1424}$	$\frac{1}{1425}$	$\frac{1}{1426}$	$\frac{1}{1427}$	$\frac{1}{1428}$	$\frac{1}{1429}$	$\frac{1}{1430}$

22. In 7713 livres, 9 to the $, how many dollars? *Ans.* $837.

23. In 648 dollars, of 8¼ livres each, how many livres? *Ans.* 5346 livres.

24. In $864, of 9 livres each, how many livres? *Ans.* 7776 livres.

The Spaniards keep their accounts in dollars and rials, 8 rials to the $, old plate, and 10 rials to the $, new plate, in some places in the West Indies.

FOREIGN EXCHANGE.

Money was originally invented to obviate the inconvenience of transporting commodities to distant places or countries, and to facilitate the distribution of the necessaries of life. Bills of exchange are still more conveniently transmitted than specie; and while the credit of a nation, or of the mercantile class of a nation, is sound, her bills will be received as payment of her foreign debts, and she is relieved from the necessity of exporting specie.

The Par of Exchange, *par pro pari*, or value for value, is determined by comparing the coins of one country with those of another, with respect to the weight of pure gold or silver they contain. But as every sovereign power possesses the right of coining money, and of altering and debasing such coins, at pleasure, hardly any two nations are found to agree in the purity and weight of their coined pieces. Hence foreign moneys are not a legal tender in the payment of domestic debts.

But as the invoices of imported goods are drawn out in the moneys of their respective countries, and the duties on many articles are charged *ad valorem*—that is, according to the value, as stated in the invoices—governments prescribe the rates at which such moneys must be estimated, and these rates pass for the par of exchange.

1431	1432	1433	1434	1435	1436	1437	1438	1439	1440
$\frac{1}{1431}$	$\frac{1}{1432}$	$\frac{1}{1433}$	$\frac{1}{1434}$	$\frac{1}{1435}$	$\frac{1}{1436}$	$\frac{1}{1437}$	$\frac{1}{1438}$	$\frac{1}{1439}$	$\frac{1}{1440}$

The Course of Exchange is that variableness in the value of foreign Bills of Exchange, which arises from the credit of the country upon which they are drawn, or from the demand in which they may happen to be. When the draughts on a nation are below par, it indicates the declension of commerce, or that her merchants are already in debt, and should send out specie rather than bills. On the contrary, when the balance of trade is in favor of a nation, and it becomes necessary to pay for her exports partly in cash, then will her bills be much in demand, and rise in their value above par.

Real moneys are actual coins ; as eagles, dollars, sovereigns, shillings, &c. Nominal or imaginary moneys, are such as have no specie representations in existence ; but which had been formerly so much in use, that they are still used in accounts, and have a name among the people. Of this kind are the £ and the marc banco.

TABLE OF GOLD COINS.

Value in Federal Money.

Coin	Value
Eagle, American	$10.00
Half Eagle, do	5.00
Half Eagle, do., 1798 and 1833	5.25
Quarter Eagle, do	2.50
Doubloon, Spanish	15.60 to 16.75
Half Doubloon, do.	7.80 to 8.37
Quarter do. do.	3.90 to 4.12
Eighth do. do.	1.90
Doubloon, Columbian	15.50 to 15.75
Eighth do	1.87
Doubloon, Mexican	15.50 to 15.75
Doubloon, New Granadian	15.50 to 15.75
Doubloon of Ecuador	15.50 to 15.75
Half Doubloon, Central American	7.75
Quarter Doubloon, Peruvian	3.87
Half Joe, Portugal (by w't.)	7.90 to 8.50
Moidore, do. do	4.70 to 6.40
Sovereign, English, 1844	4.83
Dragon Sovereign, do., 1824	4.80
Half Sovereign, do.	2.41
Guinea, do.	5.00
Half Guinea, do.	2.50
One third Guinea, do.	1.66
One Mohur, East Indies	6.75
Double Louis d'or, France	9.00
Louis d'or, do.	4.50
Forty Francs, do	7.66
Twenty Francs, do	3.83
Hundred Livre, Sardinia.	19.15
Twenty Livre, do	3.83
Ten Scudi, Rome	10.00
Twenty Lire, Italy	3.83
Quadruple Ducat, Austria	8.80
Sovereign, do	6.50

Coin	Value
Five Roubles, Russia.	$3.90
Double Frederick d'or, Prussia	7.80
Double Christian d'or, Denmark	7.80
Ten Thalers, Hanover	7.80
Five Thalers, do.	3.90
Two and a half Thalers, Hanover.	1.95
Ten Thalers, Saxony	7.80
Ten Guilders, Netherlands	4.00
Five Guilders, do	2.00
Ducat do.	2.20
Carl, Wirtemberg	4.73
Carl, Bavaria	4.73
Max, do.	3 15
Carl, Brunswick.	3.75
Ducat, Denmark.	1.72
Double Sovereign, Flanders	6.35
Pistole, Genevá	3.23
Sequin, Genou	2.16
Ducat, Germany	2.14
Georges, Hanover	3.77
Guilder, do	1.60
Ryder, Holland	5.68
Sequin, Piedmont	2.17
Frederick, Prussia	3.77
Sequin, Rome	2.12
Imperial, Russia	8.08
August, Saxony	3.72
Tical, Siam	9.05
Onza, Sicily	2.48
Adolphus, Sweden	2.66
Ruspono, Tuscany	6.50
Sequin, Venice	2.17
Five Sovereign Piece, English	24.20

1441	1442	1443	1444	1445	1446	1447	1448	1449	1450
$\frac{1}{1441}$	$\frac{1}{1442}$	$\frac{1}{1443}$	$\frac{1}{1444}$	$\frac{1}{1445}$	$\frac{1}{1446}$	$\frac{1}{1447}$	$\frac{1}{1448}$	$\frac{1}{1449}$	$\frac{1}{1450}$

Value of Silver Coins in Federal Money.

Dollar, American, - - - - $1.00
Halves and quarters in proportion.
Dimes, - - - - - - - - - .10
Half Dimes, - - - - - - - .05
Dollar, Spanish, Mexican, and Peruvian, - - - - - - 1.00
Halves, Quarters, Eighths, and Sixteenths, in proportion
Dollar, Brazil, - - - - - 1.00
Four Reals of La Plata, - - .35
Head Pistareen, - - - - - .18
Cross Pistareen, - - - - - .16
English Crown, - - - - - 1.15
English Half Crown, - - - .57
Bank Token, 3 shillings English, - - - - - - - - .50
Rupee, East Indies, - - - .40
British Colonial Quarter Dollar, - - - - - - - - - .23
English Shilling, - - - - .23
English Sixpence, - - - - .11
English Fourpence, - - - - .07
English Threepence, - - - .05
Tenpence Irish, - - - - - .12
French Crown, - - - - - 1.07
French Half Crown, - - - .50
Five Francs, French, - - - .93
Two Francs, do. - - - - .35
One Franc, do. - - - - .17
Half Franc, do. - - - - .08
Quarter Franc, do. - - - - .04
Scudo, Sicily, - - - - - - .93
Five Livre, Italy, - - - - .93
Two Livre, do. - - - - - .35
One Livre, do. - - - - - .17
Five Lire, Sardinia, - - - .93
One Livre, do. - - - - .17
Florin, Westphalia, - - - - .48
Florin, Brunswick and Lunenburg, - - - - - - - .48
Florin, Tuscany, - - - - .20
Florin, Hanover, - - - - $.50
Double Thaler, Baden, - - 1.32
Crown Thaler, do. - - - 1.04
Thaler of Baden and Hanover, .66
Thaler of Prussia, - - - - - .66
Double Thaler of Prussia, - 1.32
Imperial Thaler of Austria, - .97
Rouble, Russia, - - - - - .65
Crown Dollar of Bavaria, - - 1.04
Double Guilder of do. - - .72
German Crown, - - - - - 1.04
Crown Thaler, Hesse, - - - 1.04
Guilder of Nassau, - - - - .36
Third of a Thaler, - - - - .20
Quarter Florin, Netherlands, .08
Thirty-six Grotes, Bremen, - .30
Six Grotes, Hanse Towns, - .04
Specie Dollar, Norway, - - 1.04
Specie Dollar, Sweden, - - 1.04
Specie Rix Dollar, Denmark, 1.04
Larin, Arabia, - - - - - - .15½
Reichsthaler, Basil, - - - - 1.03
Patagon, Berne, - - - - - .94
Rupee, Bombay, - - - - - .47
Ducatoon, Flanders, - - - 1.20
Ecu, 1726, France, - - - - 1.10
Scudo, Venice, - - - - - 1.21
Patagon, Geneva, - - - - .93
Genovina, Genoa, - - - - 1.52
Old Rix Dollar, Germany, - 1.08
New do. do. - - - .97
Rix Dollar Banco, Hamburgh, 1.08
Ducatoon, Holland, - - - - 1.27
Three Florin Piece, do. - - 1.20
Rupee, Madras, - - - - - .48
Rupee, Pondicherry, - - - .46
Tymple, Poland, - - - - - .12
Scudo, Rome, - - - - - 1.00
Ducatoon, Sweden, - - - 1.20
Carolin, do. - - - - - - .30
Francescons, Tuscany, - - 1.04
Ducat, Venice, - - - - - .78

There is no uniformity in the fineness of coins, the old are generally the purest. The value of silver coins varies from 64½ to 129 cts. an ounce. The finest is the florin of Brunswick; the Tuscan 10 and 5 livre pieces are worth 124½ cts. The British standard is 120 cts. an ounce; the French, 118 cts.; the U. States, 116 cts.; the kreutzer of Austria is worth only 64½ cts. an ounce, and the dollar of New Granada 87 cts. The weight of coins is no criterion of their

1451	1452	1453	1454	1455	1456	1457	1458	1459	1460
$\frac{1}{1451}$	$\frac{1}{1452}$	$\frac{1}{1453}$	$\frac{1}{1454}$	$\frac{1}{1455}$	$\frac{1}{1456}$	$\frac{1}{1457}$	$\frac{1}{1458}$	$\frac{1}{1459}$	$\frac{1}{1460}$

value. The mass of gold is supposed to be divided into 24 parts, called carats; some pieces contain 23½ carats pure, and ½ a carat of alloy; others are as low as 18 carats. Jewelry and trinkets are seldom above 12 or 14. Notice will be taken in Alligation of the comparative purity and weight of coins.

Custom House Prices of Moneys, established by Congress.

Specie Dollar, Norway, Sweden, Denmark, - - - -	$1.05
Rix Dollar, Denmark, - -	1.00
Florin, Germany, - - - - -	.40
Florin, Augsburgh and Austria, - - - - - - - - -	.48½
Florin or Guilder, Netherlands, - - - - - - - -	.40
Thaler, Germany, Prussia, -	.69
Thaler, Bremen, - - - - -	.78¾
Livre, Tuscany and Lombardy, - - - - - - - - -	.16
Ducat, Naples, - - - - - -	.80
Ounce, Sicily, - - - - - -	2.40
£, Nova Scotia, New Brunswick, Canada, and Newfoundland, - - - - - -	$4.00
£, Great Britain, - - - -	4.84
Livre, Sardinia, France, and Belgium, - - - - - - -	.186
Rial, Vellon, Spain, - - -	.05
Marc Banco, Hamburgh, -	.35
Rouble, Russia, - - - - - -	.75
Rupee, British India, - - -	.44½
Sicca Rupee, - - - - - -	.50
Milrea, Portugal, - - - -	1.12
do. Madeira, - - - -	1.00
do. Azores, - - - - - -	.83⅓
Tale, China, - - - - - - -	1.48
Pagoda, India, - - - - - -	1.84

Bills of Exchange are orders addressed to persons, or companies, directing the payment of stipulated sums, on sight, at usance, or at times specified therein.

The parties concerned in a bill of exchange, are the *drawer*, the *drawee*, the *remitter* and the *payee*.

The drawer makes the bill and sells it; the *buyer remits* it to the drawee, in whose favor it was drawn; the drawee presents it to the *debtor*, who becomes the *acceptor*, by indorsing it with his name and the time of payment; the drawee receives the amount when due, until which time he is the *holder*, unless he pass it to another by indorsement.

The laws respecting indorsing, accepting, protesting for non-acceptance, or non-payment, and recovering by legal suit, for Bills of Exchange, require a more ample field of discussion than this volume can embrace.

Illustration of the transaction and form of a Bill of Exchange.

Mrs. B of Washington held a claim against B of Glasgow; D of Glasgow held a claim against C and R of

1461	1462	1463	1464	1465	1466	1467	1468	1469	1470
$\frac{1}{1461}$	$\frac{1}{1462}$	$\frac{1}{1463}$	$\frac{1}{1464}$	$\frac{1}{1465}$	$\frac{1}{1466}$	$\frac{1}{1467}$	$\frac{1}{1468}$	$\frac{1}{1469}$	$\frac{1}{1470}$

Washington. C and R paid Mrs B's claim, and remitted her draught on B, in favor of D, who accepted it, paid it at the time appointed, and thereby satisfied the several parties.

In class: which of these parties was the drawer? which was the remitter? which was the drawee? which was the acceptor, and the payee?

Washington, Dec. 14th, 1847.

Exchange for $1500:

At thirty days sight of this, my first of Exchange, (second and third not paid,) pay John Kent, or order, fifteen hundred dollars, with, or without further advice from me.

SAMUEL HAMILTON, ESQ.,
Cavendish Square,
London. } THOMAS BLAGDEN.

Tables of the moneys of account of different Countries.

In Great Britain and Ireland accounts are kept in pounds, shillings, pence: but the English money, until 1816, was of more value than the Irish. In that year parliament abolished that distinction. The table of this money has been already given.

The banco and current money of the Netherlands are different. The banks of Amsterdam and Hamburgh afford so much facility and security to commerce that their money is esteemed from 20 to 25 per cent. better than the current. The difference is called Agio.

At Hamburgh accounts are kept in marks, schillings and pfennigs.

12 pfennigs =1 schilling.
16 schillings =1 mark.
3 marks =1 current or rix dollar.

At Amsterdam accounts are kept in £ s. d. Flemish; but more generally in guilders, or florins, stivers, grotes, and pfennigs.

8 pfennigs =1 grote.	2 cts. U. S. =1 stiver.
2 grotes =1 stiver.	40 cents =1 guilder.
20 stivers =1 florin.	therefore
6 florins =1 £ Flemish.	240 cts =1 £ Flemish.

1471	1472	1473	1474	1475	1476	1477	1478	1479	1480
$\frac{1}{1471}$	$\frac{1}{1472}$	$\frac{1}{1473}$	$\frac{1}{1474}$	$\frac{1}{1475}$	$\frac{1}{1476}$	$\frac{1}{1477}$	$\frac{1}{1478}$	$\frac{1}{1479}$	$\frac{1}{1480}$

In France accounts are kept in francs, deciemes and centimes; or in livres, sols, and deniers.

10 centimes =1 decieme.	12 deniers =1 sol.
10 deciemes =1 franc.	20 sols=1 livre.
80 francs =81 livres.	1 livre=18 cts. 6 m.

In Spain accounts are kept in piastres, rials and maravedies, old plate. Also in dollars, rials, and quartiles; and in piastres, sols and deniers. There are as many kinds of small money as there are provinces in Spain; but the exchange is generally on the ducat of 375 maravedies.

34 maravedies=1 rial.	16 quartiles=1 rial.
8 rials=1piastre.	8 rials=1 dollar.
Also 12 deniers=1 sol.	10 cents=1 rial (plate).
20 sols=1 piastre.	5 cents=1 rial (vellon).

The new plate dollar is 10 rials.

In Portugal accounts are kept in milreas and reas. The milrea, as its name imports=1000 reas.

20 reas=1 vintin.	480 reas=1 crusade.
5 vintins=1 festoon.	1 milrea=$1.12.

At Genoa and Leghorn accounts are kept in piastres (pezo or dol.), solidi and denari. Also, in liras, solidi and denari, which are only $\frac{1}{5}$ of the value of the former.

12 denari=1 solidi.	12 denari=1 solidi.
20 solidi=1 piastre.	20 solidi=1 lira.

At Venice, in livres, sols and deniers, as in France: and in ducats and grosi. At Naples, in grains, carlins and ducats. The course of exchange is from 40 to 50 pence sterling upon the ducat banco.

12 deniers=1 sol.	10 grains=1 carlin.
20 sols=1 livre.	10 carlins=1 ducat.
24 grosi=1 ducat.	3 ducats=1 onza.

The ducat is 80 cents; the ounce is $2.40.

In Denmark accounts are kept in rix dollars and skillings; but two more denominations are included in the scale, which is as follows:

2 rustics=1 skilling.	6 marks=1 rix dollar.
16 skillings=1 mark.	1 rix dollar=1 U. S. dollar in the calculation of *ad valorem* duties here.

1481	1482	1483	1484	1485	1486	1487	1488	1489	1490
$\frac{1}{1481}$	$\frac{1}{1482}$	$\frac{1}{1483}$	$\frac{1}{1484}$	$\frac{1}{1485}$	$\frac{1}{1486}$	$\frac{1}{1487}$	$\frac{1}{1488}$	$\frac{1}{1489}$	$\frac{1}{1490}$

In Russia accounts are kept in roubles and copecks. 100 copecks=1 rouble, or 75 cents.

In Calcutta accounts are kept in rupees, annas and pice. At Madras their moneys are pagodas, fanams and cash.

12 pice=1 anna. 80 cash=1 fanam.
16 annas=1 rupee. 36 fanams=1 pagoda.

The rupee is recognized by the revenue laws of the U. S. as 44½ cents; the sicca rupee as 50 cts; the pagoda as $1.84.

The Chinese reckon in tales, mace, candarins and cash.

10 cash=1 candarin. 10 mace=1 tale.
10 candarins=1 mace. 1 tale=$1.48.

Rule.

Reduce the lower denominations to decimal parts of the superior, or to that denomination upon which the exchange is made. The decimal form of Federal money makes such reduction necessary, not only in other moneys, but also in weights and measures of every kind.

Examples for the Slate.

1. Reduce 78£ 12s. 6d. Irish or £s sterling, and reverse the same by division, at the rate of $4.84 per £ sterling.

```
12) 6.0  d.
20)12.50 s.
   78.625£
     4.84
  -------
   314500
  629000
 314500
 --------
$380.54500 Ans.
```

```
4.84)380.545(78.625 £
     3388··     20
     ----     ------
      4174    12.500 s.
      3872        12
      ----      ----
       3025    6.0 d.
       2904
       ----
        1210
         968
        ----
        2420
        2420
        ----
```

1491	1492	1493	1494	1495	1496	1497	1498	1499	1500
$\frac{1}{1491}$	$\frac{1}{1492}$	$\frac{1}{1493}$	$\frac{1}{1494}$	$\frac{1}{1495}$	$\frac{1}{1496}$	$\frac{1}{1497}$	$\frac{1}{1498}$	$\frac{1}{1499}$	$\frac{1}{1500}$

2. Reduce 48 current dollars, 2 marks, 12 schillings, 9.6 pfennigs to dollars banco of Hamburgh; agio at 20 per cent.

12) 9.6 pf.
16)12.8 sch.
3) 2.8 marks.
48.9$\frac{1}{3}$ current dollars.
$\frac{20}{100}=\frac{1}{5}$ 9.7$\frac{13}{15}$ agio.

Ans. 39.1$\frac{7}{15}$ dols. banco.

3. How many £s Flemish current are equal to 742 florins, 18 stivers, 8 pfennigs banco of Amsterdam, agio at 16$\frac{2}{3}$ per cent.?

16) 8.0 pf.
20) 18.50 stiv.
6)742.925 flor.
123.820$\frac{5}{6}$ £ Fl. banco.
$\frac{16\frac{2}{3}}{100}=\frac{1}{6}$ + 20.636$\frac{29}{36}$ agio.

Ans. 144.457$\frac{23}{36}$ £ Fl. current.

4. In 3756 francs of France, how many livres, 80 francs being equal to 81 livres? *Ans.* 3802 liv. 19 sols.

5. The money of Spain is of two kinds, plate and vellon; the vellon is to the plate as 17 to 32; then how much vellon is equal to 4725 American dollars, exchange at par on the piastre plate?

17 : 32 : : 4725 : *Ans.*

6. Lisbon remits to Boston 785 milreas, 400 reas; what sum is received at Boston, exchange at $1.12 per milrea? *Ans.* $879.648.

7. Philadelphia draws on Genoa for $1564, exchange at 97$\frac{1}{2}$ cents per pezo; how much lira money should this sum be estimated at in Genoa? *Ans.* 8020$\frac{20}{39}$.

8. Leghorn remits to New York 7898 liras 7s. 6d., at 100 cents per pezo; how much is received at New York? *Ans.* $1579.675.

9. Copenhagen draws on London for 7250 rix dollars,

1501	1502	1503	1504	1505	1506	1507	1508	1509	1510
$\frac{1}{1501}$	$\frac{1}{1502}$	$\frac{1}{1503}$	$\frac{1}{1504}$	$\frac{1}{1505}$	$\frac{1}{1506}$	$\frac{1}{1507}$	$\frac{1}{1508}$	$\frac{1}{1509}$	$\frac{1}{1510}$

at 4 rix dollars 84 skillings ($4\frac{7}{8}$) per £ sterling; what sum in sterling must be charged to Denmark when this bill is discharged? *Ans.* £1487$\frac{7}{39}$.

10. Reduce 4284 roubles 75 copecks to dollars, at 66 cents per rouble. *Ans.* $2827.935.

11. What sum is T. Law's draft worth in London, viz., $2000, exchange at 4\frac{4}{9}$ per £ sterling? *Ans.* £450.

12. How many guineas at $5 each are $12560 equal to? *Ans.* 2512 guineas.

13. How many dollars are 2348 guineas equivalent to, if 3 guineas equal 14$? *Ans.* 10957.33\frac{1}{3}$.

14. Reduce 86£ 15s. sterling to dollars, at the old par of 4\frac{4}{9}$ per £ sterling. *Ans.* 385.55\frac{5}{9}$.

15. Reduce 722£ 16s. Irish to $, at $4.47 per £ Irish. *Ans.* $3230.916.

16. In $1000 how many £s Irish, exchange at 4\frac{4}{9}$ per £? *Ans.* 225£.

Note. The old par ($40 = £9) of the sterling is still retained in use in Great Britain.

17. In $1500 how many £s, each 4\frac{4}{9}$? *Ans.* 337.5£.

18. How many dollars are equal to 337£ 10s., exchange at 4$\frac{4}{9}$ dollars per £? *Ans.* $1500.

19. In 540£ 17s. 6d. Flemish, how many rix dollars, the £ being 2$\frac{1}{2}$ rix dollars? *Ans.* $1352.1875.

20. The £ Flemish is equal to $\frac{9}{16}$ of the £ sterling; how much Flemish then, is equal to £302 5s. 8$\frac{1}{2}$d. sterling? *Ans.* £537 7s. 11$\frac{1}{9}$d.

21. Amsterdam draws on New York, for $2704.375; how much Flemish will equal this draught, if 1£=2\frac{1}{2}$? *Ans.* £1081 15s.

22. New York draws on Amsterdam for $1638, at 37$\frac{1}{2}$ cts. per florin; how much passes to the debit of New York? *Ans.* £728 Fl.

23. Baltimore draws on Amsterdam for $1250; how many guilders pass to the debit of Baltimore when this bill is accepted, exchange at 40 cents per guilder? *Ans.* 3125 guilders.

24. In 5469 livres how many francs? *Ans.* 5401 francs 48$\frac{4}{27}$ centimes.

1511	1512	1513	1514	1515	1516	1517	1518	1519	1520
$\frac{1}{1511}$	$\frac{1}{1512}$	$\frac{1}{1513}$	$\frac{1}{1514}$	$\frac{1}{1515}$	$\frac{1}{1516}$	$\frac{1}{1517}$	$\frac{1}{1518}$	$\frac{1}{1519}$	$\frac{1}{1520}$

25. Paris on Amsterdam, 5535 livres 2.7 sols, exchange at $27\frac{3}{4}$ stivers per ecu; what is the sum in Flemish?

20)2.70 sols.
5535.135 livres.
By reduction, page 90,

Then 3 liv. = 1 ecu, from the table,
and 1 ecu=$27\frac{3}{4}$ stivers, from the question.

Therefore, 3 liv. =$27\frac{3}{4}$ stivers, or 1 liv.=$9\frac{1}{4}$ stivers.

Hence 1 : $9\frac{1}{4}$: : 5535.135 : stivers.

Reduce the stivers to guilders and Flemish £s.

Ans. £426 13s. 4d.

26. Boston on Bourdeaux, for $3560, exchange at 18 cts. 5 m. on the livre; how many francs is this sum equivalent to? *Ans.* 19005.672 francs.

Note. The operator will find a circulating decimal in the work of ex. 26, viz.: 243, 243, 243, always resulting in the quotient, as long as it may be continued.

27. Paris on Philadelphia, 19243 livres 5 sols, at $18\frac{1}{2}$ cents per livre; what sum in Federal money is this equivalent to? *Ans.* $3560.

28. Cadiz on New York, 4447 piastres 16 maravedies vellon, exchange at par on the piastre plate; how much Federal money will pass to the debit of Cadiz when this bill is accepted? *Ans.* $2362.5.

29. London remits to Valencia £520 14s. 6d. sterling, exchange at 4s. 6d. sterling per pezo; how many pezos, sols, and deniers, are received at Valencia?

Ans. 2314p. 6s. 8d.

30. What sum in Portugal is equivalent to $1456 at $1.12 per milrea? *Ans.* 1300 milreas.

31. Reduce 2205 milreas 619 reas to sterling, at 5s. 3d. sterling per milrea. *Ans.* £578 19s. 6d.

32. Reduce £578 19s. 6d. sterling to milreas of 5s. 3d. sterling each. *Ans.* 2205.619 milreas.

33. Philadelphia draws on Genoa for $1564, at $97\frac{1}{2}$ cents per piastre; how much lira money is equivalent to this draft? *Ans.* $8020\frac{20}{39}$ liras.

34. Leghorn remits to New York 7898 liras 7 sols 6

1521	1522	1523	1524	1525	1526	1527	1528	1529	1530
$\frac{1}{1521}$	$\frac{1}{1522}$	$\frac{1}{1523}$	$\frac{1}{1524}$	$\frac{1}{1525}$	$\frac{1}{1526}$	$\frac{1}{1527}$	$\frac{1}{1528}$	$\frac{1}{1529}$	$\frac{1}{1530}$

deniers at $1 per pezo; how much U. S. money is received at New York? *Ans.* $1579.675.

35. Venice draws on Baltimore for 7564 ducats 18 grosi banco, exchange at 45 pence sterling per ducat, and 54d. per $, what is the sum in U. S. money?

24)18.0 grosi.
By reduction, 7564.75 ducats, page 90.

Then, since 1 ducat =45 pence,
and 54 pence= 1$,

compound, 54 : 45, or 6 : 5 is the ratio.
Therefore, 6 : 5 : : 7564.75 : x.
and $\frac{5}{6}$ comp. $\frac{1}{6}$ 1260.79+ subtract. { Case 3, Practice.

$6303.96 *Ans.*

36. Reduce 336 ducats 75 grains to dollars, the dollar being equal to 125 grains of Naples. *Ans.* $269.4.

37. London remits to Copenhagen £725 sterling, at 5 rix dollars per £ sterling; what is the sum?
Ans. 3625 rix dols.

38. Reduce 4284 roubles 75 copecks to dollars, exchange at 75 cents on the rouble. *Ans.* 3213.56\frac{1}{4}$.

39. Reduce 1260 rupees 12 annas 6 pice to $, at 50 cents on the rupee?

12) 6.0 pice.
16)12.50000 annas.
2)1260.78125 rupees.
Ans. 630.390625$

40. Reduce 480 pagodas 25 fanams 16 cash to dollars, at $1.84 per pagoda.

80) 16.0 cash.
36) 25.2 fanams.
480.7 pagodas.
× 1.84

Ans.

1531	1532	1533	1534	1535	1536	1537	1538	1539	1540
$\frac{1}{1531}$	$\frac{1}{1532}$	$\frac{1}{1533}$	$\frac{1}{1534}$	$\frac{1}{1535}$	$\frac{1}{1536}$	$\frac{1}{1537}$	$\frac{1}{1538}$	$\frac{1}{1539}$	$\frac{1}{1540}$

Questions on the Table of Silver Coins.

1. If we had the number of cents which make 1 dollar, $\frac{1}{2}$ dollar and $\frac{1}{4}$ dollar in a line, we could count them, or add them together, and find a number equal to them all; what should it be? What should the number be if there were two of each? if there were 10 of each? if there were 576 of each?

2. How many 1\$, $\frac{1}{2}$\$, and $\frac{1}{4}$\$, and of each an equal number, are in 350 cents? in 1750 cents? in 100800 cts.?

3. How many cents are equal to 1 dollar of Brazil, 1 four reals of La Plata, 1 head pistareen, and 1 cross pistareen? How many cents would two of each make? three of each? eight of each? twelve of each? 64 of each, 512 of each?

4. How many Brazilian dollars, four reals of La Plata, head pistareens, cross pistareens, and of each an equal number, would 169 cents make? would 338 cents make? would 507 cents make? would 1352 cents make? would 2028 cents make? would 10816 cents make?

5. How many cents are equal to 1 English crown, half-crown, a three shilling bank token, a rupee, a colonial quarter, one shilling, sixpence, 4 pence, 3 pence, and one Irish tenpence? How many cents would 2 of each make? 4 of each? 8 of each? 16 of each? 32 of each? 64 of each? 128 of each?

6. How many English crowns, $\frac{1}{2}$ crowns, 3 shilling bank tokens, rupees, colonial quarters, shillings, sixpences, fourpences, 3 pences, and Irish tenpences, and of each an equal number, are in 343 cents? in 686 cents? in 1372 cents? in 2744 cents? in 5488 cents? in 10976 cents? in 21952 cents? in 43904 cents?

7. How many cents are equal to 1 French crown, $\frac{1}{2}$ crown, 5 francs, 2 francs, 1 franc, $\frac{1}{2}$ franc, $\frac{1}{4}$ franc? How many cents are equal to three of each? 9 of each? 27 of each? 81 of each?

8. How many French crowns, half crowns, 5 francs, 2 francs, 1 franc, $\frac{1}{2}$ franc, and $\frac{1}{4}$ franc, and of each an equal number, are in 314 cents? 942 cents? 2826 cents? 8478 cents? 25434 cents?

1541	1542	1543	1544	1545	1546	1547	1548	1549	1550
$\frac{1}{1541}$	$\frac{1}{1542}$	$\frac{1}{1543}$	$\frac{1}{1544}$	$\frac{1}{1545}$	$\frac{1}{1546}$	$\frac{1}{1547}$	$\frac{1}{1548}$	$\frac{1}{1549}$	$\frac{1}{1550}$

9. How many cents are equal to 1 scudo of Sicily, 5 livre of Italy, 2 livre do., 1 livre do., 5 lira Sardina, 1 livre do.? how many cents are equal to 4 of each? 16 of each? 64 of each? 256 of each?

10. How many scudi of Sicily, 5 livres of Italy, 2 livres of do., 1 livre of do., 5 liras Sardinia, 1 livre do., and of each an equal number, are in 348 cents? in 1392 cents? in 5568 cents? in 22272 cents? in 89088 cents?

11. How many cents are equal to one florin of Westphalia, one florin of Brunswick and Lunenburgh, one florin of Tuscany, one florin of Hanover, one double thaler of Baden, one crown thaler do., one thaler of Baden and Hanover? How many cents are equal to 5 of each? to 25 of each? to 125 of each?

12. How many florins and thalers, as enumerated in example 11, and of each an equal number, are in 468 cents? in 2340 cents? in 11700 cents? in 58500 cents?

13. How many cents are equal to 1 thaler and 1 double thaler of Prussia, 1 imperial thaler of Austria, 1 rouble of Russia, 1 crown dollar and 1 double guilder of Bavaria, 1 German crown, 1 crown thaler of Hesse, and 1 guilder of Nassau? how many cents are in 6 of each? in 36 of each? in 216 of each?

14. How many of the coins enumerated in ex. 13, and of each an equal number, are in 780 cents? in 4680 cents? in 28080 cents? in 168480 cents?

15. How many cents are equal to $\frac{1}{4}$ florin, Netherlands; 36 grotes, Bremen; 6 grotes, Hanse Towns; specie dollar of Norway, Sweden, or Denmark; larin, of Arabia; reichsthaler, Basil; patagon, Berne and Geneva; rupee, Bombay; ducatoon, Flanders; ecu, France; scudo, Venice; genovina, Genoa? how many cents are equal to 8 of each? 64 of each? 512 of each?

16. How many of the coins enumerated in example 15, and of each an equal number, are in 1001$\frac{1}{2}$ cents? in 8012 cents? in 64096 cents? in 512768 cents?

17. How many cents are equal to 1 old rix dollar, Germany; 1 new do.; 1 rix dollar banco, Hamburgh; 1 ducatoon, Holland; 1 three florin piece, do.; 1 rupee,

1551	1552	1553	1554	1555	1556	1557	1558	1559	1560
$\frac{1}{1551}$	$\frac{1}{1552}$	$\frac{1}{1553}$	$\frac{1}{1554}$	$\frac{1}{1555}$	$\frac{1}{1556}$	$\frac{1}{1557}$	$\frac{1}{1558}$	$\frac{1}{1559}$	$\frac{1}{1560}$

Madras; 1 rupee, Pondicherry; 1 tymple, Poland; 1 scudo, Rome; 1 ducatoon and 1 carolin, Sweden; 1 francescono, Tuscany; 1 ducat, Venice? How many cents are equal to 9 of each? 81 of each? 729 of each?

18. How many of the coins enumerated in example 17, and of each an equal number, are equal to 1098 cents? are in 9882 cents? in 88938 cents? in 800442 cents?

Questions on the Table of Gold Coins.

19. How many cents are equal to 1 of each of the first twenty coins in the table of gold coins? how many cents are equal to 1 of each of the second twenty? how many cents are equal to 1 of each of the third twenty? how many cents are equal to one of each of the last seven? how many cents are equal to 1 of each of the whole 67 gold coins mentioned in the table?

20. How many of the first twenty coins in the table of gold coins, and of each an equal number, are in 15732 cents? how many of the second 20, and of each an equal number, in 12616 cents? how many of the third 20, number as before, are in 7515 cents? of the last 7 in 5078 cents? how many of the whole 67 coins enumerated in the table of gold coins, and of each an equal number, are in 40941 cents?

Compound Exchanges, or Circular Remittances.

This rule applies to an extensive foreign commerce; in which a merchant sitting in his office, and having funds abroad calculates from the range of his correspondence and from the state of the foreign markets, the advantage he may have by transferring his claim from one country to another, and another, instead of bringing home his funds by a direct remittance, or importation of goods.

RULE. Following the route of the intended exchange ascertain by single comparisons, how much of each kind of money, along the route, the funds to be transferred will make: you will at last find how much the amount will be when brought home.

1561	1562	1563	1564	1565	1566	1567	1568	1569	1570
$\frac{1}{1561}$	$\frac{1}{1562}$	$\frac{1}{1563}$	$\frac{1}{1564}$	$\frac{1}{1565}$	$\frac{1}{1566}$	$\frac{1}{1567}$	$\frac{1}{1568}$	$\frac{1}{1569}$	$\frac{1}{1570}$

The difference between a direct and a circular remittance will be the gain or loss; or,

Express the rates of exchange by equations as follows:

3 A=4 B.
5 B=6 C.
4 C=9 D; of which the terms on the left are antecedents, those on the right are consequents; the sum to be exchanged is the unmatched term, for which an equivalent is to be found.

Multiply the given sum by the consequents, and divide the product by the antecedents, continually; the result will be the sum to be brought home, with all its increase, or decrease.

This operation affords the best opportunity for canceling terms that are common to both sides of the equations, or reducing to their lowest terms such couplets as have a common measure.

Examples for the Slate.

1. Received from Paris an account of sales, of my 400 barrels of flour, the neat proceeds amounting to 17200 livres, which, by my order is sent to Cadiz, at 5 livres per piastre; from Cadiz it is sent to Lisbon, at 6 piastres for 5½ milreas, and from Lisbon to New York, at 1 milrea for $1.12; what have I gained by this circular remittance, the livre being valued here at 18½ cents?

liv.		pia.		liv.		pia.	
5	:	1	: :	17200	:	3440	
pia.		mil.		pia.		mil.	
6	:	5½	: :	3440	:	3153⅓	
mil.		$		mil.		$	
1	:	1.12	:	3153⅓	:	3531.73⅓	cir. rem.
liv.		cts.		liv.		$	
1	:	18½	:	17200	:	3182	direct rem.

Ans. $349.73⅓ gained.

Note. (2d term×3d term)÷1st=4th term.

1571	1572	1573	1574	1575	1576	1577	1578	1579	1580
$\frac{1}{1571}$	$\frac{1}{1572}$	$\frac{1}{1573}$	$\frac{1}{1574}$	$\frac{1}{1575}$	$\frac{1}{1576}$	$\frac{1}{1577}$	$\frac{1}{1578}$	$\frac{1}{1579}$	$\frac{1}{1580}$

Or thus, liv. 5=1 pia.
pia. 6=5½ mil.
mil. 1=1.12 dol.
dol. ? =17200 liv.

3,0)10595,2

3531.73⅓ circ. remittance.
17200×18½= 3182. direct remittance.

Ans. $349.73⅓ gained.

2. Leghorn remits my $100 to Venice at $95 to 100 ducats banco; Venice remits the proceeds to Cadiz at 1 ducat for 350 maravedies; Cadiz remits the proceeds to Lisbon at 272 maravedies to 630 reas; Lisbon sends them to Amsterdam at 400 reas to 4 shillings Flemish; Amsterdam sends them to Philadelphia at £1 Flemish to $2½: to how much do these several exchanges increase my $100? *Ans.* $106.67.

3. How much will £200 Flemish be worth in the United States, after exchanging on Paris at 4½ shillings Flemish for 3 livres; on London at 1 livre for 10 pence sterling, on Dublin at 12d. sterling for 13 pence Irish, and on Baltimore at £1 Irish for $4⁴⁄₉? *Ans.* $534.98.

4. If 6 English crowns equal 7 Brazilian dollars, and 35 B. dollars equal 100 four real pieces of La Plata, and 80 four real pieces L. P. equal 25 milreas of Portugal, and 100 milreas of Portugal equal 112 milreas of Madeira, and 83⅓ milreas of Madeira equal 100 milreas of the Azores, and 3 milreas of the Azores equal ½ English guinea, and 9 guineas equal 10 French louis d'ors, and 3 louis d'ors equal 2 mohurs of India, and 4 mohurs equal 13 five guilder pieces, and 1 five guilder piece equal $2; how many $s are 100 English crowns worth, after passing through so many exchanges?

Ans. $112.34+

Note. Foreign weights and measures may be arbitrated after the same manner.

1581	1582	1583	1584	1585	1586	1587	1588	1589	1590
$\frac{1}{1581}$	$\frac{1}{1582}$	$\frac{1}{1583}$	$\frac{1}{1584}$	$\frac{1}{1585}$	$\frac{1}{1586}$	$\frac{1}{1587}$	$\frac{1}{1588}$	$\frac{1}{1589}$	$\frac{1}{1590}$

For suppose 4 lbs. of A are worth 3 lbs. of B, and 5 lbs. of B worth 4 lbs. of C, and 6 lbs. of C worth 5 lbs. of D, then how many lbs. of D are worth 120 lbs. of A?

A ~~4~~= 3 B
B ~~5~~= ~~4~~ C
C 6= ~~5~~ D
D ?=120 A

6)360

Ans. 60 lbs.

EXCHANGE OF COMMODITIES, OR BARTER.

This is the most ancient species of commerce; the exchange of one article for another must have been practised at the first formation of society. This kind of traffic ceased with the introduction of money, and every article required its valuation in cash before any exchange could be effected. Hence in every calculation of this kind two ranks of proportionals are to be considered: one to ascertain how much money the quantity of one of the articles will come to, the other to find how much of the other article the money will buy. This is but an application of the rules of calculation already laid down.

Examples for the Slate.

1. How much sugar at 12 cents per lb. must be delivered for 20 cwt. of tobacco at $10.8 per cwt.?

cwt.		$		cwt.		$
1	:	10.8	: :	20	:	216
$		lb.		$		lbs.
.12	:	1	: :	216	:	1800

And $1800 \div 112 = 16\frac{8}{112}$, or 16 cwt. 8 lbs. *Ans.*

1591	1592	1593	1594	1595	1596	1597	1598	1599	1600
$\frac{1}{1591}$	$\frac{1}{1592}$	$\frac{1}{1593}$	$\frac{1}{1594}$	$\frac{1}{1595}$	$\frac{1}{1596}$	$\frac{1}{1597}$	$\frac{1}{1598}$	$\frac{1}{1599}$	$\frac{1}{1600}$

2. A barters 10 pieces of calico, each 25 yards at $18\frac{1}{2}$ cents per yard with B, for pepper at 15 cents per lb.; what quantity of pepper must A call for?

Ans. $308\frac{1}{3}$ lbs.

3. If 250 yards of calico at $18\frac{1}{2}$ cents per yard be given in exchange for $308\frac{1}{3}$ lbs. of pepper, at what price per lb. is the pepper rated? *Ans.* at 15 cents.

4. B barters 1000 yards of linen at 38 cents per yard with C, for flannel at 41 cents per yard; how much flannel should B receive? *Ans.* $926\frac{34}{41}$.

5. C has brandy at 80 cents per gallon, D has 126 yards of cloth at \$1.2 per yard; how much brandy should C give to D for his cloth? *Ans.* 189 gallons.

6. D has silk at 14s. per lb., B has cloth at 10s. per yard in cash, which in barter he puts at 12s. 6d.; how must D's silk be rated to make his profits equal to B's?

Ans. at 17s. 6d. per lb.

7. E has linen worth 20d. per ell, ready money, but in barter he will have 2s.; B has cloth worth 14.5s. per yard in cash; how should B's cloth be rated in barter?

Ans. 17.4s.

8. G has cloth which he barters with L at 10 cts. per yard more than it cost him, against tea at 12.5s. per lb. L paid 10s. per lb. for the tea; required the prime cost of the cloth? *Ans.* 3s. 4d. per yard.

9. H has cloth at \$5.5 per yard; how many yards of this must he give in exchange for 1760 yds. of B's linen, at 45 cents per yard? *Ans.* 144 yards.

10. L labors 120 days for M, at .9\$ per day, for which he receives corn at .75\$ per bushel; what quantity of corn does the labor of L amount to? *Ans.* 144 bushels.

11. P has 144 reams of paper at \$1.6 per ream, which he barters with S for 216 pairs of shoes; at what price does S rate his shoes by the pair? *Ans.* at $\$1.06\frac{2}{3}$.

12. H has 41 cwt. of hops, for which Q delivers to him 17 cwt. 3 qrs. 4 lbs. of prunes, at 10 cts. per lb., and in cash \$96. Query, how much are the hops rated at per cwt.? *Ans.* \$7.2.

1601	1602	1603	1604	1605	1606	1607	1608	1609	1610
$\frac{1}{1601}$	$\frac{1}{1602}$	$\frac{1}{1603}$	$\frac{1}{1604}$	$\frac{1}{1605}$	$\frac{1}{1605}$	$\frac{1}{1607}$	$\frac{1}{1608}$	$\frac{1}{1609}$	$\frac{1}{1610}$

GAINS AND LOSSES.

RULE. To find the gain or loss on any transaction, compare the cost and charges with the returns it makes. If the given amount involve gain or loss per cent., then 100 : 100+rate of gain, or 100 : 100—rate of loss, as the given amount is to the same amount +, or —, the whole gain or loss. The variations of this proportion will unfold all the cases, which can occur when the gain or loss is at a certain rate per cent.

Examples for the Slate.

1. If 980 lbs. of merchandise are bought for $588, and sold for $666.4, what is the gain per lb.?

```
        $666.4
         588.
        ------
980)    78.40(.08 dols. Ans.
        78 40
        =====
```

2. Bought 53 yards of cloth at .6£ per yard, and sold the same at 14s. per yard; what was the whole gain?
Ans. 5.3£.

3. Bought 12 hhds. of wine at $40 per hhd., paid sundry charges $15, and sold the same for $49.5. per hhd.; how much was the gain on this transaction? *Ans.* $99.

4. Sold 18 cwt. of cheese for $252, by which I gained 25 per cent.; how much was the prime cost per lb.?
Ans. 10 cts.

5. At 7½ cts. pr. lb., what is the profit on 1 cwt.?
Ans. $8.40.

6. If 1 lb. of cloves cost $2¼, and is sold for $2.16 per lb.; how much is the loss per cent.? *Ans.* 4 per cent.

7. If the cloves above mentioned were sold for $2.34, how much would be the gain per cent.?
Ans. 4 per cent.

1611	1612	1613	1614	1615	1616	1617	1618	1619	1620
$\frac{1}{1611}$	$\frac{1}{1612}$	$\frac{1}{1613}$	$\frac{1}{1614}$	$\frac{1}{1615}$	$\frac{1}{1616}$	$\frac{1}{1617}$	$\frac{1}{1618}$	$\frac{1}{1619}$	$\frac{1}{1620}$

8. A miller bought 1000 bushels of wheat at 10 shillings per bushel, Maryland currency; he paid for transporting and other charges £10; he afterward ground it into flour, making 200 barrels, which he sold at \$7$\frac{1}{2}$ per barrel; how much was his gain per cent.?

Ans. 10$\frac{5}{17}$ per cent.

9. At 16 per cent. loss, what is the proportional loss on 1 lb. of sugar, sold at 10$\frac{1}{2}$ pence? *Ans.* 2 pence.

10. Sold 53 yards of cloth at .7£ per yard, by which was gained £5. 6s.; what did it cost per yard?

Ans. 12 shillings.

11. Sold 12 hogsheads of wine for \$594 by which my gain is \$99; how much did it cost per gallon?

Ans. 65$\frac{10}{21}$ cts.

12. Sold 12 yards of cloth for \$96, by which was gained 10 per cent.; what was the prime cost per yard?

Ans. \$7.27+

13. For how much must 1 cwt. of coffee which cost \$27 be sold, that the gain may be 10 per cent.?

Ans. \$29.7.

14. For how much must 1 cwt. of coffee be sold which cost \$27, that the loss may be 10 per cent.?

Ans. \$24.3.

15. Bought 1 tun of wine for \$250 ready money, and sold it again for \$272.5 payable in 8 months; how much is the gain, discount being computed at 6 per cent. per annum? *Ans.* \$12$\frac{2}{104}$.

16. Received from Belfast 1760 yards Irish linen, which cost there 18d. Irish per yard; the government premium upon the exportation of linen is 1d. to the shilling; this is to be deducted from the cost; the freight and charges amounted to \$50, to be added to the cost; how must this linen be sold per yard to gain 30 per cent., the Irish £ being \$4$\frac{4}{9}$. *Ans.* \$.434$\frac{61}{396}$.

17. Bought 50 cwt. of butter at the rate of \$2 for 12 lbs., in ready money; and sold the same again for 25 cents per lb., payable in 6 months; discounting at 6 per cent. per annum, how much is the neat gain?

Ans. \$425.89.

1621	1622	1623	1624	1625	1626	1627	1628	1629	1630
$\frac{1}{1621}$	$\frac{1}{1622}$	$\frac{1}{1623}$	$\frac{1}{1624}$	$\frac{1}{1625}$	$\frac{1}{1626}$	$\frac{1}{1627}$	$\frac{1}{1628}$	$\frac{1}{1629}$	$\frac{1}{1630}$

DIVISION OF JOINT STOCKS.

This rule, which is usually called Fellowship, teaches to divide gains and losses among partners in trade, in the ratio of their respective stocks, or in the ratio composed of the ratios of their stocks and times. It applies to the division of Bank, Insurance, Railroad, Canal and all other stocks, whether the scale be great or small.

Case 1. When the stocks in company are considered without reference to time.

RULE. Say, as the sum of all the stocks is to the whole gain or loss, so is each partner's stock to his share of the gain or loss.

Proof. Add all the gains or losses into one sum; this reverses the rule.

Divide 6 in the ratio of 1 and 2

1+2 is to 6 as { 1 is to 2. / 2 is to 4. } *Ans*

In the same manner divide

15 in the ratio of 2 and 3.
21 in the ratio of 3 and 4.
27 in the ratio of 4 and 5.
33 in the ratio of 5 and 6.
39 in the ratio of 6 and 7.
45 in the ratio of 7 and 8.
51 in the ratio of 8 and 9.
57 in the ratio of 9 and 10.
63 in the ratio of 10 and 11.
69 in the ratio of 11 and 12.
49 in the ratio of 2 and 5.
50 in the ratio of 3 and 7.
52 in the ratio of 4 and 9.
48 in the ratio of 5 and 11.
60 in the ratio of 7 and 13.
63 in the ratio of 8 and 13.
81 in the ratio of 13 and 14.
96 in the ratio of 15 and 17.

Divide 12 in the ratio of 1, 2 and 3.

1+2+3 is to 12 as { 1 is to 2. / 2 is to 4. / 3 is to 6. } *Ans.*

Proof 12

1631	1632	1633	1634	1635	1636	1637	1638	1639	1640
$\frac{1}{1631}$	$\frac{1}{1632}$	$\frac{1}{1633}$	$\frac{1}{1624}$	$\frac{1}{1635}$	$\frac{1}{1636}$	$\frac{1}{1637}$	$\frac{1}{1638}$	$\frac{1}{1639}$	$\frac{1}{1640}$

In like manner divide the following numbers in the ratio of the annexed figures.

27 as 2, 3 and 4.
48 as 3, 4 and 5.
60 as 4, 5 and 6.
54 as 5, 6 and 7.
63 as 6, 7 and 8.
72 as 7, 8 and 9.
81 as 8, 9 and 10.
90 as 9, 10 and 11.
99 as 10, 11 and 12.
108 as 11, 12 and 13.
117 as 12, 13 and 14.
126 as 13, 14 and 15.
135 as 14, 15 and 16.
144 as 15, 16 and 17.

Examples for the Slate.

1. A and B lay out in goods £80, which they sell again for £100; A's stock was £30, B's stock £50; what was each man's gain in this transaction?

As 80 : 20 : : { 30 : £7 10s. A's gain. / 50 : £12 10s. B's gain. } *Ans.*

2. B and C purchase goods in company. B's stock is $450 dols. and C's 600 dols., they gain 273 dols. Query the proportional gain of each?

Ans. B's $117, C's $156.

3. A debtor is owing to C $500, to D $900; but his whole estate amounts to no more than $1100; what do his creditors severally lose by him?

Ans. C loses 107\frac{1}{7}$: D 192\frac{6}{7}$.

4. F, G, and H, freight a ship with 108 tuns of wine from Madeira; F's share was 48, G's 36, and H's 24 tuns; a storm arising, the seamen threw 45 tuns overboard; how much should each merchant sustain of this loss? *Ans.* F 20, G 15, and H 10 tuns.

5. The sum of 450 dols. was divided among three men, A had $\frac{1}{3}$, B 100 dols.; what was the share of C?

Ans. 200 dols.

6. Three butchers pay for a lot of cattle 3000 dols., Q paid as much again as P, and R paid as much as Q and P, they gain 25 per cent; what is the proportional gain of each? *Ans.* P's $125, Q's $250, R's $375.

7. Two merchants gained $450, of which L is to have 3 times as much as T; how much is each to have?

Ans. L $337.5, T $112.5.

1641	1642	1643	1644	1645	1646	1647	1648	1649	1650
$\frac{1}{1641}$	$\frac{1}{1642}$	$\frac{1}{1643}$	$\frac{1}{1644}$	$\frac{1}{1645}$	$\frac{1}{1646}$	$\frac{1}{1647}$	$\frac{1}{1648}$	$\frac{1}{1649}$	$\frac{1}{1650}$

8. Let $1000 be divided among 3 persons, A to have $\frac{2}{3}$ as much as B, and C as much as both; what will be the quota of each? *Ans.* C $500, B $300, A $200.

9. A, B, and C, put in money together; A put in £20, B £30, C a sum unknown; they gained £36, C claims thereof £16; what did A and B gain, and C put in?
Ans. A gained £8, B £12, and C put in £40.

10. A, B, and C put in money together, A put in £20, B and C put in £85, they gain $\frac{3}{5}$ of their whole stock; B took up £21; how much did A and C gain, and B and C put in?
Ans. A gained £12, C £30, B put in £35, C £50.

11. Divide $5250 between 3 sons, A, B, and C, so that A may have $\frac{4}{9}$ of the whole, B $\frac{3}{5}$ of the remainder, and C the residue, and tell the share of each?
Ans. A's 2333\frac{1}{3}$, B's $1750, C's 1166\frac{2}{3}$.

Note. The shares are as 4, 3, and 2. For B's part is $\frac{3}{5}$ of $\frac{5}{9}=\frac{3}{9}$, and consequently C's $\frac{2}{9}$ of the whole; thus, with the same denominators, as $\frac{4}{9}$, $\frac{3}{9}$, and $\frac{2}{9}$, fractions are to one another in the ratios of their numerators.

12. C, M, P, and G engaged to build a house for $6231, of which C received $\frac{12}{31}$ parts of the whole, M $\frac{5}{6}$, and P $\frac{1}{2}$ as much as C; G's share was $603; how much were the shares of C, M, and P?
Ans. C $2412, M $2010, and P $1206.

Case 2. When stocks remain in trade to unequal measures of time.

Rule. Multiply each particular stock by its time; then, the sum of the products is to the whole gain or loss, as each particular product is to its share of the gain or loss.

Because each product is such a stock as will bear in one year or one month, the same gain or loss which each stock would bear in the years or months used as a multiplier.

Example. 14 will bear in 1 month as much as 7 will bear in 2 months; 24 will bear in 1 month as much as 8 will bear in 3 months.

1651	1652	1653	1654	1655	1656	1657	1658	1659	1660
$\frac{1}{1651}$	$\frac{1}{1652}$	$\frac{1}{1653}$	$\frac{1}{1654}$	$\frac{1}{1655}$	$\frac{1}{1656}$	$\frac{1}{1657}$	$\frac{1}{1658}$	$\frac{1}{1659}$	$\frac{1}{1660}$

Examples for the Slate.

1. Divide 114 in the ratio composed of the ratios of 2 to 3, and of 7 to 8.

thus : $2\times7=14$
$3\times8=24$

$$\overline{38} : 114 : : \begin{cases} 14 : 42. \\ 24 : 72. \end{cases} \textit{Ans.}$$

Proof 114

In like manner divide

2. 129 in the ratio of 3×5 to 4×7. *Ans.* 45 and 84.
3. 232 in the ratio of 5×8 to 6×3. *Ans.* 160 and 72.
4. 264 in the ratio of 9×4 to 5×6. *Ans.* 144 and 120.
5. 340 in the ratio of 11×3 to 7×5. *Ans.* 165 and 175.
6. 339 in the ratio of 4×8 to 9×9. *Ans.* 96 and 243.
7. 356 in the ratio of 6×9 to 5×7. *Ans.* 216 and 140.
8. 170 in the ratio of 8×2 to 3×6. *Ans.* 80 and 90.
9. 960 in the ratio of 12×5 to 4×9. *Ans.* 600 and 360.

10. A and B trade in company, A put in 250 dols. for 4 months, B 150 dols. for 8 months ; they gain 275 dols. ; required to know the particular gain of each ?

$ mos.
A 250×4=1000
B 150×8=1200

$$\text{As } \overline{2200} : 275 : : \begin{cases} 1000 : 125 \text{ A's.} \\ 1200 : 150 \text{ B's.} \end{cases} \textit{Ans.}$$

11. A, R, and T in a joint adventure put in as follows : A 200 dols. for 4 months, R 150 dols. for 6 months, T 125 dols. for 8 months ; they gain 540 dols. ; required the quota of each ?

Ans. A's 160 dols., R's 180 dols., T's 200 dols.

12. Three graziers hired a pasturage for 145.2 dols., A put in 5 oxen for $4\frac{1}{2}$ months, B put in 8 oxen for 5 months, and C 9 oxen for $6\frac{1}{2}$ months ; how much should each man pay ?

Ans. A 27 dols., B 48 dols., and C 70.2 dols.

1661	1662	1663	1664	1665	1666	1667	1668	1669	1670
$\frac{1}{1661}$	$\frac{1}{1662}$	$\frac{1}{1663}$	$\frac{1}{1664}$	$\frac{1}{1665}$	$\frac{1}{1666}$	$\frac{1}{1667}$	$\frac{1}{1668}$	$\frac{1}{1669}$	$\frac{1}{1670}$

13. R, S, and T join stocks, R's stock is £4000 for 12 months, S's £3000 for 15 months, T's £5000 for 8 months; they gain £665; what is each man's particular share thereof? *Ans.* R's £240, S's £225, T's £200.

14. L, M, and N join stocks, L puts in 400 yds. of linen for 12 months, M 200 dols. cash for 10 months, N 300 pairs of shoes for 8 months; they gain 298 dols., of which M takes 100, and N 90; required the price of L's linen per yard, and of N's shoes per pair?

Ans. Linen 45 cts., shoes 75 cts.

15. A, C, and E make a common stock for 12 mos., A puts in £125, 2 months after C put in £125, and 2 months after that E puts in £150; at the end of 6 months A takes out £25, 2 months after that C takes out £25, and E £30, the whole gain is £179; what part thereof should each man have? *Ans.* A £67.5, C £57.5, and E £54.

VULGAR FRACTIONS.

A single part is a fraction whose numerator is 1.

A plural part is a fraction whose numerator is 2 or more.

Every single part is the first term of a series of fractions which have a common denominator.

Proper fractions are such as have their numerators less than their denominators.

Improper fractions are such as have their numerators equal to or greater than their denominators.

A simple fraction contains only one expression.

A compound fraction is a part of another part; thus 1d. is $\frac{1}{12}$ of $\frac{1}{20}$ of 1£.

A mixed quantity is a multiple and a part connected.

A continued fraction has a mixed quantity for one of its terms.

In the following table the fractions on the left margin are single parts: all the rest are plural parts.

1671	1672	1673	1674	1675	1676	1677	1678	1679	1680
$\frac{1}{1671}$	$\frac{1}{1672}$	$\frac{1}{1673}$	$\frac{1}{1674}$	$\frac{1}{1675}$	$\frac{1}{1676}$	$\frac{1}{1677}$	$\frac{1}{1678}$	$\frac{1}{1679}$	$\frac{1}{1680}$

Every line across the page is a series of fractions which have a common denominator.

The fractions on the left and below the diagonal line are proper fractions. Those on the right and above are improper fractions; and each of these which range along the line is equal to 1.

All the examples given in the table are of simple fractions.

Table of Proper and Improper Fractions.

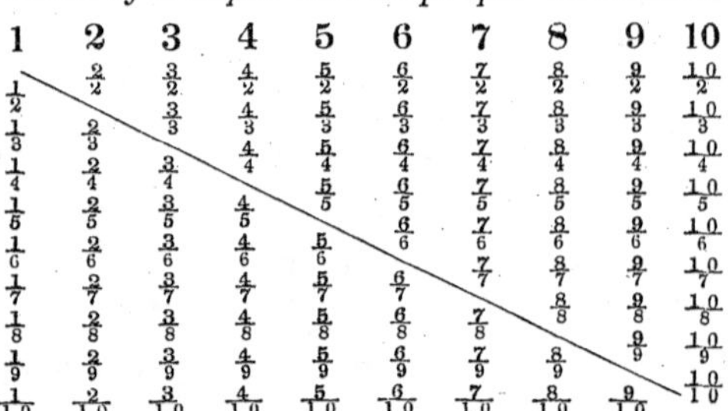

1	2	3	4	5	6	7	8	9	10
	$\frac{2}{2}$	$\frac{3}{2}$	$\frac{4}{2}$	$\frac{5}{2}$	$\frac{6}{2}$	$\frac{7}{2}$	$\frac{8}{2}$	$\frac{9}{2}$	$\frac{10}{2}$
$\frac{1}{2}$		$\frac{3}{3}$	$\frac{4}{3}$	$\frac{5}{3}$	$\frac{6}{3}$	$\frac{7}{3}$	$\frac{8}{3}$	$\frac{9}{3}$	$\frac{10}{3}$
$\frac{1}{3}$	$\frac{2}{3}$		$\frac{4}{4}$	$\frac{5}{4}$	$\frac{6}{4}$	$\frac{7}{4}$	$\frac{8}{4}$	$\frac{9}{4}$	$\frac{10}{4}$
$\frac{1}{4}$	$\frac{2}{4}$	$\frac{3}{4}$		$\frac{5}{5}$	$\frac{6}{5}$	$\frac{7}{5}$	$\frac{8}{5}$	$\frac{9}{5}$	$\frac{10}{5}$
$\frac{1}{5}$	$\frac{2}{5}$	$\frac{3}{5}$	$\frac{4}{5}$		$\frac{6}{6}$	$\frac{7}{6}$	$\frac{8}{6}$	$\frac{9}{6}$	$\frac{10}{6}$
$\frac{1}{6}$	$\frac{2}{6}$	$\frac{3}{6}$	$\frac{4}{6}$	$\frac{5}{6}$		$\frac{7}{7}$	$\frac{8}{7}$	$\frac{9}{7}$	$\frac{10}{7}$
$\frac{1}{7}$	$\frac{2}{7}$	$\frac{3}{7}$	$\frac{4}{7}$	$\frac{5}{7}$	$\frac{6}{7}$		$\frac{8}{8}$	$\frac{9}{8}$	$\frac{10}{8}$
$\frac{1}{8}$	$\frac{2}{8}$	$\frac{3}{8}$	$\frac{4}{8}$	$\frac{5}{8}$	$\frac{6}{8}$	$\frac{7}{8}$		$\frac{9}{9}$	$\frac{10}{9}$
$\frac{1}{9}$	$\frac{2}{9}$	$\frac{3}{9}$	$\frac{4}{9}$	$\frac{5}{9}$	$\frac{6}{9}$	$\frac{7}{9}$	$\frac{8}{9}$		$\frac{10}{10}$
$\frac{1}{10}$	$\frac{2}{10}$	$\frac{3}{10}$	$\frac{4}{10}$	$\frac{5}{10}$	$\frac{6}{10}$	$\frac{7}{10}$	$\frac{8}{10}$	$\frac{9}{10}$	

A Lemma. To find the least common multiple of several numbers:

RULE. While any two or more given numbers have a common measure divide them by it, bringing down to a new line the prime numbers and quotients; the continual product of the divisors, quotients, and prime numbers will be the least common multiple. The design of this lemma is to prepare common denominators for fractions, without which they cannot be added or subtracted.

Examples for the Slate.

1. What is the least common multiple of 3, 4 and 6? also of 3, 5, 7, and 9?

```
3) 3  4  6              3) 3  5  7  9
   -------                 ----------
2) 1  4  2                 1  5  7  3
   -------                 ----------
   1  2  1                 3× 5× 7× 3=315 Ans.
   -------
   3× 2× 2=12 Ans.
```

1681	1682	1683	1684	1685	1686	1687	1688	1689	1690
$\frac{1}{1681}$	$\frac{1}{1682}$	$\frac{1}{1683}$	$\frac{1}{1684}$	$\frac{1}{1685}$	$\frac{1}{1686}$	$\frac{1}{1687}$	$\frac{1}{1688}$	$\frac{1}{1689}$	$\frac{1}{1690}$

Note. When one of the given terms is a measure of another of them, the measure may be omitted, for the rule might be thus: find the least common multiple of two of the given terms, and also the least common multiple of the common multiple just found and a third, and so on: but as in the examples above, 6 is a common multiple of itself and 3; and 9 is also a common multiple of itself and 3. Hence 3 might have been omitted in both.

2. What is the least common multiple of 4, 6, 8 and 10? also of 4, 7, 8 and 12? *Ans.* 120 and 168.

3. What is the least number divisible by each of the nine digits? or what is the least common multiple of 1, 2, 3, 4, 5, 6, 7, 8, 9, 10?

Case 1. To reduce a fraction to its least terms.

RULE. Divide both the terms of the fraction by their greatest common measure; the quotients express the fraction in its least terms. See page 94.

Or, divide the terms of the fraction by any common measure, and these quotients by any other common measure, and so on, till they become prime to each other, in which form they express the fraction in its lowest numbers.

Examples for the Slate.

1. Reduce $\frac{216}{288}$ to its lowest terms.

```
216)288(1
    216
    ---
     72)216(3
        216
        ===
```

$\frac{216}{288} \div 72 = \frac{3}{4}$. *Ans.*

Or thus: $\frac{216}{288} \div 8 = \frac{27}{36} \div 9 = \frac{3}{4}$. *Ans.*

2. Reduce $\frac{195}{780}$ to its least terms. *Ans.* $\frac{1}{4}$.
3. Reduce $\frac{136}{204}$ to its least terms. *Ans.* $\frac{2}{3}$.
4. Reduce $\frac{525}{630}$ to its least terms. *Ans.* $\frac{5}{6}$.

The division of both terms of a fraction by the same number makes no change in its value. See page 38.

Note 1. 2 will divide any even number.

2. 5 will divide any number ending with 5 or a cipher.

3. An equal number of 0s may be rejected from the right of both terms of a fraction; for this is to divide by 10, 100, &c.

4. If 4 divide two figures on the right of any number, it will divide the number; because 100 is a multiple of 4.

1691	1692	1693	1694	1695	1696	1697	1698	1699	1700
$\frac{1}{1691}$	$\frac{1}{1692}$	$\frac{1}{1693}$	$\frac{1}{1694}$	$\frac{1}{1695}$	$\frac{1}{1696}$	$\frac{1}{1697}$	$\frac{1}{1698}$	$\frac{1}{1699}$	$\frac{1}{1700}$

5. If 8 divide three figures on the right of a number, it will divide the number; because 1000 is a multiple of 8.

6. If 9 or 3 divide the sum of the digits of any number, 9 or 3 will divide the number.

7. If the sum of the digits of the odd ranks, as the 1st, 3d, 5th, &c., be equal to the sum of the digits of the even ranks, as the 2d, 4th, 6th, &c., then 11 will divide such number.

Case 2. To reduce any number to a fraction of a given denominator.

RULE. Subscribe a unit for a denominator, and multiply the terms of this fraction by the given denominator.

Examples for the Slate.

1. Reduce 7 to a fraction whose denominator shall be 9; and 15 to a fraction whose denominator shall be 11.

$\frac{7}{1}\times9=\frac{63}{9}$ *Ans.* and $\frac{15}{1}\times11=\frac{165}{11}$ *Ans.*

2. Reduce 31 to a fraction whose denominator shall be 15; and 48 to a fraction whose denominator shall be 18. *Ans.* $\frac{465}{15}$, and $\frac{864}{18}$.

3. Reduce 45 to a fraction whose denominator shall be 12, and 72 to a fraction whose denominator shall be 16. *Ans.* $\frac{540}{12}$, and $\frac{1152}{16}$.

Case 3. To reduce a quantity composed of a multiple and a part or parts of a unit to an improper fraction.

RULE. Subscribe a unit for a denominator, then multiply the terms of this new fraction by the denominator of the given part or parts.

Examples for the Slate.

1. Reduce $24\frac{7}{8}$ to an improper fraction.

$$\frac{24\frac{7}{8}}{1}\times8=\frac{199}{8}\quad \textit{Ans.}$$

Note. To multiply a fraction by its own denominator is to remove the denominator, leaving the numerator as the product.

2. Reduce $12\frac{1}{2}$ and $18\frac{3}{4}$ to improper fractions. *Ans.* $\frac{25}{2}$ and $\frac{75}{4}$.

3. Reduce $31\frac{1}{4}$ and $37\frac{1}{2}$ to improper fractions. *Ans.* $\frac{125}{4}$ and $\frac{75}{2}$.

1701	1702	1703	1704	1705	1706	1707	1708	1709	1710
$\frac{1}{1701}$	$\frac{1}{1702}$	$\frac{1}{1703}$	$\frac{1}{1704}$	$\frac{1}{1705}$	$\frac{1}{1706}$	$\frac{1}{1707}$	$\frac{1}{1708}$	$\frac{1}{1709}$	$\frac{1}{1710}$

4. Reduce $14\frac{3}{8}$ and $15\frac{9}{16}$ to improper fractions. *Ans.* $\frac{115}{8}$ and $\frac{249}{16}$.
5. Reduce $144\frac{25}{30}$ to an improper fraction. *Ans.* $\frac{4345}{30}$.
6. Reduce $183\frac{5}{21}$ to an improper fraction. *Ans.* $\frac{3848}{21}$.
7. Reduce $45\frac{7}{8}$ and $64\frac{11}{16}$ to improper fractions. *Ans.* $\frac{367}{8}$ and $\frac{1035}{16}$.

Case 4. To reduce an improper fraction to a whole or mixed number; which reverses the last case.

Rule. Divide the numerator by the denominator, and if there be a remainder, write the divisor under it.

Examples for the Slate.

1. Reduce $\frac{199}{8}$ to a mixed number, and $\frac{12}{3}$ to a whole number.

```
 199                       12
 ---                       --
  8 )199(24 7/8 Ans.        3)12
     16                       --
     --                        4 Ans.
      39                      ====
      32             Or thus, 12/3=4. Ans.
      --
       7
      ==
```

It is unnecessary to make a long division when the divisor is small.

2. Reduce $\frac{25}{2}$ and $\frac{75}{4}$ to mixed numbers.
3. Reduce $\frac{125}{4}$ and $\frac{75}{2}$ to mixed numbers.
4. Reduce $\frac{115}{8}$ and $\frac{249}{16}$ to mixed numbers.
5. Reduce $\frac{4345}{30}$, $\frac{367}{8}$, and $\frac{1035}{16}$ to mixed numbers.

See the answers of these given in the examples of Case 3.

Case 5. To reduce a compound fraction to a simple one.

Rule. Multiply all the numerators together for the numerator, and all the denominators for the denominator of the simple fraction.

Note 1. If any of the simple given fractions be a whole or mixed number, reduce it to an improper fraction.
2. Reduce the simple fractions to their least terms.
3. Cancel terms that are common.

1711	1712	1713	1714	1715	1716	1717	1718	1719	1720
$\frac{1}{1711}$	$\frac{1}{1712}$	$\frac{1}{1713}$	$\frac{1}{1714}$	$\frac{1}{1715}$	$\frac{1}{1716}$	$\frac{1}{1717}$	$\frac{1}{1718}$	$\frac{1}{1719}$	$\frac{1}{1720}$

Examples for the Slate.

1. Reduce $\frac{1}{2}$ of $\frac{2}{3}$ to a simple fraction; also $\frac{1}{3}$ of $\frac{3}{4}$: or, let $\frac{1}{2}$ of $\frac{2}{3}$ of $\frac{3}{4}$ be reduced to a simple fraction.

Thus, $\frac{1}{2}\times\frac{2}{3}=\frac{1}{3}$, $\frac{1}{3}\times\frac{3}{4}=\frac{1}{4}$, canceling 2s and 3s.

or, $\frac{1}{2}\times\frac{2}{3}\times\frac{3}{4}=\frac{6}{24}$, multiplying the terms without canceling.

2. Reduce $\frac{2}{3}$ of $\frac{3}{5}$ of $\frac{10}{11}$ to a simple fraction. *Ans.* $\frac{4}{11}$.

3. Reduce $\frac{5}{8}$ of $\frac{4}{5}$ of $\frac{2}{3}$ to a simple fraction. *Ans.* $\frac{1}{3}$.

4. Reduce $\frac{9}{11}$ of $\frac{4}{5}$ of $\frac{11}{12}$ to a simple fraction. *Ans.* $\frac{3}{5}$.

5. Reduce $\frac{2}{5}$ of $\frac{5}{8}$ of $3\frac{1}{2}$ to a simple fraction. *Ans.* $\frac{7}{8}$.

6. Reduce $\frac{2}{5}$ of $\frac{9}{16}$ of 100 to a simple fraction. *Ans.* $\frac{45}{2}$.

7. Reduce $\frac{7}{8}$ of $\frac{8}{9}$ of $\frac{9}{16}$ to a simple fraction. *Ans.* $\frac{7}{16}$.

8. Reduce $\frac{37}{2}$ of $\frac{4}{100}$ to a simple fraction. *Ans.* $\frac{37}{50}$.

9. Reduce $\frac{12}{2}$ of $\frac{4}{75}$ to a simple fraction. *Ans.* $\frac{8}{25}$.

10. Reduce $43\frac{3}{4}$ of $\frac{7}{8}$ of $\frac{1}{100}$ to a simple fraction. *Ans.* $\frac{49}{128}$.

Case 6. To reduce fractions of different denominators to equivalent fractions having a common denominator.

RULE 1. Reduce the given fractions to their simplest forms, then multiply the numerator and denominator of each by all the other denominators. (See page 38.)

2. Make the least common multiple of all the denominators the common denominator, then each of the given denominators is to its numerator as the common denominator is to the new numerators respectively.

Examples for the Slate.

1. Reduce $\frac{2}{3}$, $\frac{3}{4}$, and $\frac{5}{6}$ to a common denominator.

$2\times4\times6=48$, numerator for $\frac{2}{3}$. *Ans.* $\frac{48}{72}$.
$3\times3\times6=54$, numerator for $\frac{3}{4}$. *Ans.* $\frac{54}{72}$.
$5\times3\times4=60$, numerator for $\frac{5}{6}$. *Ans.* $\frac{60}{72}$

$3\times4\times6=\overline{72}$, common denominator.

Note. Since 72 is not the least common multiple of the denominators 3, 4, and 6, the fractions found are not in their least terms.

1721	1722	1723	1724	1725	1726	1727	1728	1729	1730
$\frac{1}{1721}$	$\frac{1}{1722}$	$\frac{1}{1723}$	$\frac{1}{1724}$	$\frac{1}{1725}$	$\frac{1}{1726}$	$\frac{1}{1727}$	$\frac{1}{1728}$	$\frac{1}{1729}$	$\frac{1}{1730}$

Second Method.

$$\begin{array}{r} \quad 2 \;\; 3 \;\; 5 \\ 3)\overline{3} \;\; \overline{4} \;\; \overline{6} \\ \hline 2)1 \;\; 4 \;\; 2 \\ \hline 1 \;\; 2 \;\; 1 \end{array}$$

Then $3 \times 2 \times 2 = 12$.

Therefore, 3 : 2 : : 12 : 8
4 : 3 : : 12 : 9
6 : 5 : : 12 : 10

Ans. $\frac{8}{12}$, $\frac{9}{12}$, $\frac{10}{12}$.

This method expresses the given fractions in the lowest terms in which they can have a common denominator, since 12 is the least common multiple of the given denominators.

2. Reduce $\frac{2}{3}$, $\frac{3}{5}$, and $\frac{3}{4}$ to a common denominator.
Ans. $\frac{40}{60}$, $\frac{36}{60}$, and $\frac{45}{60}$.

3. Reduce $\frac{3}{8}$, $2\frac{1}{2}$, and 4 to a common denominator.
Ans. $\frac{3}{8}$, $\frac{20}{8}$, and $\frac{32}{8}$.

4. Reduce $\frac{3}{7}$ of $\frac{5}{3}$ and $\frac{3}{4}$ of $\frac{4}{5}$ to a common denominator.
Ans. $\frac{25}{35}$ and $\frac{21}{35}$.

5. Reduce 4, $\frac{15}{21}$, and $\frac{7}{9}$ of $\frac{8}{10}$ to a common denominator.
Ans. $\frac{1260}{315}$, $\frac{225}{315}$, and $\frac{196}{315}$.

6. Reduce $\frac{2}{3}$, $\frac{4}{5}$, and $\frac{5}{6}$ to a common denominator.
Ans. $\frac{20}{30}$, $\frac{24}{30}$, and $\frac{25}{30}$.

Note. In very many instances fractions may be reduced to a common denominator by simple inspection: thus, multiplying the terms of the series of halves by 2, 3, or 4, reduces it to a series of fourths, sixths, or eighths; also, multiplying the terms of the series of thirds by 2, 3, or 4, reduces it to a series of sixths, ninths, or twelfths. See the table, page 174.

Case 7. To reduce fractions of mercantile numbers from one denomination to another.

Rule. First make the given fraction compound by comparing it with all the intervening denominations, then reduce it to a simple fraction, and to its lowest terms.

Examples for the Slate.

1. Reduce $\frac{2}{15}$ of a £ to the fraction of a penny.

Thus, $\frac{2}{15}$ of $\frac{20}{1}$ of $\frac{12}{1} = \frac{480}{15}$d. But since either 3, 5, or 15 is a common measure of 480 and 15, the fraction in its lowest terms will be $\frac{32}{1}$d.

1731	1732	1733	1734	1735	1736	1737	1738	1739	1740
$\frac{1}{1731}$	$\frac{1}{1732}$	$\frac{1}{1733}$	$\frac{1}{1734}$	$\frac{1}{1735}$	$\frac{1}{1736}$	$\frac{1}{1737}$	$\frac{1}{1738}$	$\frac{1}{1739}$	$\frac{1}{1740}$

2. Reduce $\frac{1}{3}$d. to the fraction of a £.

$\frac{1}{3}$ of $\frac{1}{12}$ of $\frac{1}{20}=\frac{1}{720}$£. *Ans.*

3. Reduce $\frac{1}{2}$ of a cent to part of a $. *Ans.* $\frac{1}{200}$.
4. Reduce $\frac{3}{4}$ of a dime to parts of a $. *Ans.* $\frac{3}{40}$.
5. Reduce $\frac{5}{8}$ of a $ to part of an eagle. *Ans.* $\frac{1}{16}$.
6. Reduce $\frac{3}{4}$ of an ounce troy to part of 1 lb. *Ans.* $\frac{1}{16}$.
7. Reduce $\frac{2}{7}$ of 1 cwt. to parts of 1 qr. *Ans.* $\frac{8}{7}$.
8. Reduce $\frac{3}{4}$ of 1 cwt. to parts of 1 ton. *Ans.* $\frac{3}{80}$.
9. Reduce $\frac{3}{5}$ of 1 lb. troy to its value in dwts. *Ans.* $\frac{144}{1}$.
10. Reduce $\frac{3}{4}$ of 1 quart to parts of a gallon. *Ans.* $\frac{3}{16}$.
11. Reduce $\frac{7}{8}$ of a furlong to parts of 1 mile. *Ans.* $\frac{7}{64}$.
12. Reduce $\frac{3}{4}$ of a yard to its value in nails. *Ans.* $\frac{12}{1}$.
13. Reduce $\frac{3}{20}$ of an ounce troy to part of 1 lb. *Ans.* $\frac{1}{80}$.
14. Reduce $\frac{2}{3}$ of $\frac{3}{4}$ of 4 hours to part of 1 day. *Ans.* $\frac{1}{12}$.

Case 8. To reduce fractions of the higher titles in mercantile numbers to units of the inferior grades.

RULE. Multiply the unit of which the part is given by the numerator of the fraction, and divide the product by the denominator, as in division of mercantile numbers.

Examples for the Slate.

1. Reduce $\frac{3}{8}$ of a $ to cents and mills, and $\frac{7}{8}$ of a £ to shillings and pence.

```
 3   $ c.m.              1£= 20s.
 8) 3.00.0                    7
    ------                  ---
Ans. .37.5                8)140
                            ---
                       Ans. 17s. 6d.
```

2. Reduce $\frac{3}{8}$ of a shilling to pence. *Ans.* $4\frac{1}{2}$d.
3. Reduce $\frac{3}{16}$ $ to cents and parts. *Ans.* $18\frac{3}{4}$ cts.

1741	1742	1743	1744	1745	1746	1747	1748	1749	1750
$\frac{1}{1741}$	$\frac{1}{1742}$	$\frac{1}{1743}$	$\frac{1}{1744}$	$\frac{1}{1745}$	$\frac{1}{1746}$	$\frac{1}{1747}$	$\frac{1}{1748}$	$\frac{1}{1749}$	$\frac{1}{1750}$

4. Reduce $\frac{5}{8}$\$ to cents and part. *Ans.* $62\frac{1}{2}$ cts.
5. Reduce $\frac{3}{4}$ lb. troy to ounces. *Ans.* 9 oz.
6. Reduce $\frac{5}{16}$ cwt. to qrs. and lbs. *Ans.* 1 qr. 7 lbs.
7. Reduce $\frac{1}{12}$ mile to fur., po., yds., &c.
Ans. 7 fur. 13 po. 1 yd. $2\frac{1}{2}$ feet.
8. Reduce $\frac{3}{10}$ E. ell to qrs. and na. *Ans.* 1 qr. 2 na.
9. Reduce $\frac{7}{16}$ acre to roods and poles.
Ans. 1 R. 30 po.
10. Reduce $\frac{3}{10}$ day to hrs. min. *Ans.* 7 hrs. 12 min.
11. Reduce $\frac{4}{5}$ of 3£ 15s. to £s. *Ans.* 3£.
12. Reduce $\frac{5}{6}$ of 2 lbs. 2 oz. 2 drs. apoth. to its value.
Ans. 1 lb. 9 oz. 7 drs.
13. What is the value of $\frac{5}{9}$ of 3 hhds. 36 gals.?
Ans. 1 hhd. 62 gals.
14. What is the value of $\frac{3}{19}$ of 1 bu. 3 pe. 1 qt.?
Ans. 1 pe. 1 qt.

15. What is the value of $\frac{3}{5}$ of 5 English guineas, in sterling money at 21s. and in Federal money at \$5 each?
Ans. Sterling £3 3s. and \$15.

Case 9. To reduce the inferior grades of mercantile numbers to fractions of the superior.

RULE. Make the inferior given grades the numerator and a unit of the superior grade the denominator; then reduce both to the same name, and to their lowest terms.

Examples for the Slate.

1. Reduce 17s. 6d. to parts of £1.

$$\text{It makes } \frac{17\text{s. }6\text{d.}}{1£}=\frac{17\frac{1}{2}}{20}=\frac{35}{40}=\frac{7}{8}£. \textit{ Ans.}$$

2. Reduce $2\frac{1}{2}$ cents to the fraction of 1 dime, and $37\frac{1}{2}$ cts. to parts of \$1.

$$\text{Thus } \frac{2\frac{1}{2}}{10}=\frac{5}{20}=\frac{1}{4}, \text{ and } \frac{37\frac{1}{2}}{100}=\frac{75}{200}=\frac{3}{8}. \textit{ Ans.}$$

3. Reduce 1 qr. 14 lbs. to parts of 1 cwt. *Ans.* $\frac{3}{8}$.
4. Reduce $4\frac{1}{2}$d. to parts of 1 shilling. *Ans.* $\frac{3}{8}$.
5. Reduce $18\frac{3}{4}$ cts. to parts of \$1. *Ans.* $\frac{3}{16}$.
6. Reduce $62\frac{1}{2}$ cts. to parts of \$1. *Ans.* $\frac{5}{8}$.

1751	1752	1753	1754	1755	1756	1757	1758	1759	1760
$\frac{1}{1751}$	$\frac{1}{1752}$	$\frac{1}{1753}$	$\frac{1}{1754}$	$\frac{1}{1755}$	$\frac{1}{1756}$	$\frac{1}{1757}$	$\frac{1}{1758}$	$\frac{1}{1759}$	$\frac{1}{1760}$

7. Reduce 9 ounces to parts of 1 lb. troy. *Ans.* $\frac{3}{4}$.
8. Reduce 1 qr. 7 lbs. to parts of 1 cwt. *Ans.* $\frac{5}{16}$.
9. Reduce 1 lb. 9 oz. 7 drs. to parts of 2 lbs. 2 oz. 2 drs. apoth. *Ans.* $\frac{5}{6}$.
10. Reduce 7 fur. 13 po. 1 yd. 2 ft. 6 in. to parts of 1 mile. *Ans.* $\frac{11}{12}$.
11. Reduce 1 qr. 2 na. to parts of an E. ell. *Ans.* $\frac{3}{10}$.
12. Reduce 1 R. 13 sq. po. to parts of an acre. *Ans.* $\frac{53}{160}$.
13. Reduce 1 hhd. 62 gals. to parts of 3 hhds. 36 gals. *Ans.* $\frac{5}{9}$.
14. Reduce 1 pe. 1 qt. to parts of 1 bu. 3 pe. 1 qt. *Ans.* $\frac{3}{19}$.
15. Reduce 7 hrs. 12 minutes to parts of 1 day. *Ans.* $\frac{3}{10}$.

Addition of Vulgar Fractions.

To add two single fractions, take the sum of their denominators for numerator, and the product of the same for denominator.

Class Exercise upon the line of fractions.

$\frac{1}{1}+\frac{1}{2}=\frac{3}{2}$, $\frac{1}{2}+\frac{1}{3}=\frac{5}{6}$, $\frac{1}{3}+\frac{1}{4}=\frac{7}{12}$,
$\frac{1}{4}+\frac{1}{5}=\frac{9}{20}$, $\frac{1}{5}+\frac{1}{6}=\frac{11}{30}$, $\frac{1}{6}+\frac{1}{7}=\frac{13}{42}$,
$\frac{1}{7}+\frac{1}{8}=\frac{15}{56}$, $\frac{1}{8}+\frac{1}{9}=\frac{17}{72}$, $\frac{1}{9}+\frac{1}{10}=\frac{19}{90}$, &c.

Continue this exercise along the line of single fractions until the practice become familiar and easy.

Note 1. In finding a common denominator each numerator is multiplied by all the denominators but its own; and since, in adding single fractions, the numerator is 1, it makes no increase of the product: therefore the sum of the denominators is the same as the sum of their products ×1.

RULE. Reduce the given fractions to a common denominator, if necessary; then the sum of the new numerators will be the numerator, under which write the common denominator, for the fractional sum required.

Add first only the fractional parts of mixed numbers, and to that sum add the whole numbers.

1761	1762	1763	1764	1765	1766	1767	1768	1769	1770
$\frac{1}{1761}$	$\frac{1}{1762}$	$\frac{1}{1763}$	$\frac{1}{1764}$	$\frac{1}{1765}$	$\frac{1}{1766}$	$\frac{1}{1767}$	$\frac{1}{1768}$	$\frac{1}{1769}$	$\frac{1}{1770}$

Examples for the Slate.

1. What is the sum of $\frac{5}{12}$ and $\frac{11}{12}$?

These fractions have a common denominator, viz. 12; we therefore add the numerators, thus:

$$\frac{5}{12}+\frac{11}{12}=\frac{16}{12}=\frac{4}{3}=1\frac{1}{3}.\ \textit{Ans.}$$

2. What is the sum of $\frac{4}{5}$ and $\frac{5}{6}$?

$$\frac{4}{5}+\frac{5}{6}=\frac{24}{30}+\frac{25}{30}=\frac{49}{30}=1\frac{19}{30}\ \textit{Ans.}$$

3. What is the sum of $\frac{3}{4}$ of $\frac{5}{6}$, $7\frac{1}{2}$, $\frac{3}{4}$ and $4\frac{1}{2}$?

$$\frac{3}{4} \text{ of } \frac{5}{6}=\frac{5}{8},\ 7\frac{1}{2}+4\frac{1}{2}=12,$$

$$\text{To which add } \frac{5}{8}+\frac{3}{4}=\frac{11}{8}=\ 1\frac{3}{8}$$

$$13\frac{3}{8}\ \textit{Ans.}$$

4. What is the sum of $\frac{1}{3}$ of 28, $2\frac{1}{2}$, and $\frac{3}{8}$ of $\frac{9}{10}$? *Ans.* $12\frac{41}{240}$.
5. What is the sum of $\frac{5}{8}$ of 18, $\frac{3}{4}$, $\frac{2}{3}$, and $\frac{3}{7}$? *Ans.* $13\frac{2}{21}$.
6. What is the sum of $\frac{7}{8}$ yard and $\frac{3}{4}$ nail? *Ans.* 3 qrs. $2\frac{3}{4}$ na.
7. What is the sum of $\frac{2}{3}$, $\frac{3}{4}$, $\frac{7}{8}$ and $22\frac{1}{4}$? *Ans.* $24\frac{13}{24}$.
8. What is the sum of $21\frac{1}{4}$ yards and $33\frac{5}{12}$ Fl. ells? *Ans.* 46 yds. 1 qr. 1 na.
9. What is the sum of £$\frac{1}{8}$ $\frac{2}{3}$s. and $\frac{9}{20}$ of a crown or 5s. ster.? *Ans.* 5s. 5d.
10. What is the sum of $\frac{3}{8}$ ton and $\frac{15}{16}$ cwt.? *Ans.* 8 cwt. 1 qr. 21 lbs.
11. What is the sum of $\frac{9}{10}$ mile and $\frac{2}{11}$ fur.? *Ans.* 7 fur. 15 po. 1 yd. 1 ft. 6 in.
12. What is the sum of $\frac{8}{9}$ day and $\frac{5}{6}$ year? *Ans.* 305 days 6 hrs. 20 m.
13. What is the sum of $\frac{1}{2}$ of $\frac{2}{3}$ of $\frac{3}{4}$ dol. and $22\frac{3}{16}$ dols.? *Ans.* $22.4375.
14. What is the sum of $\frac{7}{8}$ of $9\frac{5}{6}$ and $\frac{3}{7}$ of $\frac{15}{16}$? *Ans.* $9\frac{1}{168}$.
15. What is the sum of $\frac{7}{9}$ of 5 and $\frac{4}{9}$ of $7\frac{7}{8}$? *Ans.* $7\frac{7}{8}$.

1771	1772	1773	1774	1775	1776	1777	1778	1779	1780
$\frac{1}{1771}$	$\frac{1}{1772}$	$\frac{1}{1773}$	$\frac{1}{1774}$	$\frac{1}{1775}$	$\frac{1}{1776}$	$\frac{1}{1777}$	$\frac{1}{1778}$	$\frac{1}{1779}$	$\frac{1}{1780}$

Subtraction of Vulgar Fractions.

To subtract two single fractions; take the difference of the denominators for the numerator, and their product for the denominator of the difference sought.

Class Exercise upon the line of fractions.

$\frac{1}{2}-\frac{1}{3}=\frac{1}{6}$, $\frac{1}{3}-\frac{1}{4}=\frac{1}{12}$, $\frac{1}{4}-\frac{1}{5}=\frac{1}{20}$,
$\frac{1}{5}-\frac{1}{6}=\frac{1}{30}$, $\frac{1}{6}-\frac{1}{7}=\frac{1}{42}$, $\frac{1}{7}-\frac{1}{8}=\frac{1}{56}$,
$\frac{1}{8}-\frac{1}{9}=\frac{1}{72}$, $\frac{1}{9}-\frac{1}{10}=\frac{1}{90}$, $\frac{1}{10}-\frac{1}{11}=\frac{1}{110}$, &c.

Continue this exercise along the line until the practice become familiar and easy.

RULE. Reduce the given fractions to a common denominator, if necessary; subtract the less numerator from the greater, and write the common denominator under the difference, for the answer.

Examples for the Slate.

1. From $\frac{7}{9}$ take $\frac{3}{8}$:
   thus, $\frac{7}{9}-\frac{3}{8}=\frac{56}{72}-\frac{27}{72}=\frac{29}{72}$ *Ans.*
2. From $\frac{11}{12}$ take $\frac{25}{72}$:
   thus, $\frac{11}{12}-\frac{25}{72}=\frac{66}{72}-\frac{25}{72}=\frac{41}{72}$ *Ans.*
3. From $\frac{33}{38}$ take $\frac{6}{19}$. *Ans.* $\frac{21}{38}$.
4. From $\frac{3}{4}$ of $\frac{7}{8}$ take $\frac{2}{3}$ of $\frac{9}{10}$. *Ans.* $\frac{9}{160}$.
5. From $\frac{1}{2}$ of $\frac{23}{24}$ take $\frac{1}{5}$ of $\frac{10}{16}$. *Ans.* $\frac{17}{48}$.
6. From $\frac{2}{3}$ of 56, take $\frac{1}{7}$ of 64. *Ans.* $28\frac{4}{21}$.
7. From $19\frac{1}{2}$ take $5\frac{7}{8}$. *Ans.* $13\frac{5}{8}$.
8. From $\frac{2}{3}$ of a £ take $7\frac{3}{8}$ shillings. *Ans.* 5s. $11\frac{1}{2}$d.
9. From $\frac{5}{8}$ of a cwt. take $24\frac{3}{4}$ lbs.
   Ans. 1 qr. 17 lbs. 4 oz.
10. From $\frac{2}{11}$ mile take $\frac{5}{11}$ furlong. *Ans.* 1 fur.
11. From $\frac{3}{5}$ of an English ell take $\frac{1}{4}$ yd. *Ans.* 2 qrs.
12. From 9 days take $\frac{3}{4}$ week. *Ans.* 3 days 18 hrs.
13. From $\frac{15}{16}$ dol. take $43\frac{3}{4}$ cts. *Ans.* .5 dol.
14. From £2 take $26\frac{1}{3}$ of a shilling. *Ans.* 13s. 8d.
15. From 1 eagle take $\frac{1}{2}$ of $\frac{16}{3}$ of a $. *Ans.* 7.33\frac{1}{3}$.
16. From $\frac{3}{4}$ of a lb. troy take $\frac{1}{6}$ of an ounce.
    Ans. 8 oz. 16 dwt. 16 gr.

1781	1782	1783	1784	1785	1786	1787	1788	1789	1790
$\frac{1}{1781}$	$\frac{1}{1782}$	$\frac{1}{1783}$	$\frac{1}{1784}$	$\frac{1}{1785}$	$\frac{1}{1786}$	$\frac{1}{1787}$	$\frac{1}{1788}$	$\frac{1}{1789}$	$\frac{1}{1790}$

17. From $3\frac{1}{2}$ weeks take $9\frac{7}{10}$ days.
Ans. 2 ws. 19 hrs. 12 m.

18. From $12\frac{5}{12}$ take $6\frac{1}{2}$. *Ans.* $5\frac{11}{12}$.

Questions on Addition and Subtraction of single Fractions.

1. From the sum of $\frac{1}{2}$ and $\frac{1}{3}$ take $\frac{1}{4}$. *Ans.* $\frac{7}{12}$.
2. From the sum of $\frac{1}{2}$ and $\frac{1}{4}$ take $\frac{1}{3}$. *Ans.* $\frac{5}{12}$.
3. From the sum of $\frac{1}{3}$ and $\frac{1}{4}$ take $\frac{1}{2}$. *Ans.* $\frac{1}{12}$.
4. From the sum of $\frac{1}{4}$ and $\frac{1}{5}$ take $\frac{1}{6}$. *Ans.* $\frac{17}{60}$.
5. From the sum of $\frac{1}{4}$ and $\frac{1}{6}$ take $\frac{1}{5}$. *Ans.* $\frac{13}{60}$.
6. From the sum of $\frac{1}{5}$ and $\frac{1}{6}$ take $\frac{1}{4}$. *Ans.* $\frac{7}{60}$.
7. From the sum of $\frac{1}{6}$ and $\frac{1}{7}$ take $\frac{1}{8}$. *Ans.* $\frac{31}{168}$.
8. From the sum of $\frac{1}{6}$ and $\frac{1}{8}$ take $\frac{1}{7}$. *Ans.* $\frac{25}{168}$.
9. From the sum of $\frac{1}{7}$ and $\frac{1}{8}$ take $\frac{1}{6}$. *Ans.* $\frac{17}{168}$.
10. From the sum of $\frac{1}{8}$ and $\frac{1}{9}$ take $\frac{1}{10}$. *Ans.* $\frac{49}{360}$.
11. From the sum of $\frac{1}{8}$ and $\frac{1}{10}$ take $\frac{1}{9}$. *Ans.* $\frac{41}{360}$.
12. From the sum of $\frac{1}{9}$ and $\frac{1}{10}$ take $\frac{1}{8}$. *Ans.* $\frac{31}{360}$.

Multiplication of Vulgar Fractions.

In fractions denominators are divisors; and when a fraction becomes a factor, its power as a divisor affects the product. Hence, in multiplying by fractions, the upper terms multiply and the lower terms divide: for this reason, terms that are common may be rejected or canceled; and terms which have a common measure may be divided by it, in order to abridge the work.

Rule. Reduce the given fractions to their simplest forms; then rejecting terms that are common, take the product of all the numerators for the numerator, and the product of all the denominators for the denominator of the required product.

Examples for the Slate.

1. What are the products of $\frac{1}{2}$ and $\frac{2}{3}$? $\frac{2}{3}$ and $\frac{3}{4}$? $\frac{3}{4}$ and $\frac{4}{5}$?
Ans. $\frac{1}{2}\times\frac{2}{3}=\frac{2}{6}$, or $\frac{1}{3}$. $\frac{2}{3}\times\frac{3}{4}=\frac{2}{4}$, or $\frac{1}{2}$. $\frac{3}{4}\times\frac{4}{5}=\frac{3}{5}$.

2. Required the continual product of $\frac{1}{2}$, $\frac{2}{3}$, $4\frac{1}{4}$, $3\frac{3}{4}$ and $\frac{7}{8}$?

First reduce the fractions to their simplest forms:

1791	1792	1793	1794	1795	1796	1797	1798	1799	1800
$\frac{1}{1791}$	$\frac{1}{1792}$	$\frac{1}{1793}$	$\frac{1}{1794}$	$\frac{1}{1795}$	$\frac{1}{1796}$	$\frac{1}{1797}$	$\frac{1}{1798}$	$\frac{1}{1799}$	$\frac{1}{1800}$

$\frac{1}{2}\times\frac{2}{3}\times\frac{17}{4}\times\frac{15}{4}\times\frac{7}{8}$: this may be reduced by rejecting the 2s and dividing the 15 by 3; thus, $\frac{17}{4}\times\frac{5}{4}\times\frac{7}{8}$; there is now no other reduction possible; therefore $17\times5\times7$ will be the numerator, and $4\times4\times8$ will be the denominator, and the fraction is $\frac{595}{128}$; which being an improper fraction, may be reduced to the form of a mixed number, viz.: $4\frac{83}{128}$.

3. Required the product of $\frac{2}{9}$ and $\frac{7}{10}$? *Ans.* $\frac{7}{45}$.
4. Required the product of $\frac{4}{15}$ and $\frac{5}{24}$? *Ans.* $\frac{1}{18}$.
5. Multiply $\frac{3}{7}$, $\frac{4}{11}$, $\frac{5}{14}$ and $\frac{21}{22}$. *Ans.* $\frac{45}{847}$.
6. Required the product of $\frac{5}{7}$ of $\frac{2}{3}$ and 6? *Ans.* $2\frac{6}{7}$.
7. Required the product of $9\frac{1}{2}$ and $8\frac{1}{4}$? *Ans.* $78\frac{3}{8}$.
8. Required the product of $\frac{5}{6}$ of 8 and $\frac{3}{4}$ of 7? *Ans.* 35.
9. Required the product of $18\frac{1}{3}$ and $2\frac{1}{3}$? *Ans.* $42\frac{7}{9}$.
10. Required the continual product of $\frac{1}{2}$, 2, $\frac{2}{3}$, 3, $\frac{3}{4}$, 4, $\frac{4}{5}$, 6 and $\frac{5}{6}$? *Ans.* 24.

Division of Vulgar Fractions.

When a fraction becomes a divisor its denominator multiplies; because its effect is always contrary to that of the numerator. And this multiplication is effected in two ways, viz.: either by dividing the denominator, or multiplying the numerator of the dividend; that is, by increasing the value of each part, or the number of the parts.

RULE. Reduce the given fractions to their simplest forms; then divide the numerator by the numerator, and the denominator by the denominator, if it can be done exactly; otherwise, invert the terms of the divisor and use multiplication.

Examples for the Slate.

1. Required the quotient of $\frac{1}{2}$ by $\frac{1}{3}$? of $\frac{1}{3}$ by $\frac{1}{2}$? of $\frac{1}{3}$ by $\frac{1}{4}$? of $\frac{1}{4}$ by $\frac{1}{3}$? of $\frac{1}{4}$ by $\frac{1}{5}$? of $\frac{1}{5}$ by $\frac{1}{4}$?

$\frac{1}{2}\div\frac{1}{3}=\frac{1}{2}\times\frac{3}{1}=\frac{3}{2}=1\frac{1}{2}$. $\frac{1}{3}\div\frac{1}{2}=\frac{1}{3}\times\frac{2}{1}=\frac{2}{3}$.

$\frac{1}{3}\div\frac{1}{4}=\frac{1}{3}\times\frac{4}{1}=\frac{4}{3}=1\frac{1}{3}$. $\frac{1}{4}\div\frac{1}{3}=\frac{1}{4}\times\frac{3}{1}=\frac{3}{4}$.

$\frac{1}{4}\div\frac{1}{5}=\frac{1}{4}\times\frac{5}{1}=\frac{5}{4}=1\frac{1}{4}$. $\frac{1}{5}\div\frac{1}{4}=\frac{1}{5}\times\frac{4}{1}=\frac{4}{5}$.

Carry this practice extensively along the line of single fractions.

1801	1802	1803	1804	1805	1806	1807	1808	1809	1810
$\frac{1}{1801}$	$\frac{1}{1802}$	$\frac{1}{1803}$	$\frac{1}{1804}$	$\frac{1}{1805}$	$\frac{1}{1806}$	$\frac{1}{1807}$	$\frac{1}{1808}$	$\frac{1}{1809}$	$\frac{1}{1810}$

2. Required the quotient of $\frac{32}{6} \div \frac{8}{3}$? and $\frac{8}{3} \div \frac{32}{6}$?
$\frac{32}{6} \div \frac{8}{3} = \frac{4}{2} = 2$. *Ans.* $\frac{8}{3} \div \frac{32}{6} = \frac{8}{3} \times \frac{6}{32} = \frac{48}{96} = \frac{1}{2}$. *Ans.*
3. Required the quotient of $22\frac{1}{3}$ by $7\frac{1}{3}$? *Ans.* $3\frac{1}{22}$.
4. Required the quotient of $\frac{7}{16}$ by $\frac{3}{4}$? *Ans.* $\frac{7}{12}$.
5. Required the quotient of $\frac{5}{6}$ by 3? *Ans.* $\frac{5}{18}$.
6. Required the quotient of $\frac{5}{8}$ by $4\frac{1}{2}$? *Ans.* $\frac{5}{36}$.
7. Required the quotient of $\frac{1}{2}$ of $\frac{2}{3}$ by $\frac{3}{4}$ of $\frac{5}{6}$? *Ans.* $\frac{8}{15}$.
8. Required the quotient of $3\frac{3}{4}$ by $\frac{1}{2}$ of $\frac{3}{4}$ of $2\frac{1}{2}$? *Ans.* 4.
9. Divide $\frac{1}{2}$ of $\frac{3}{4}$ of $2\frac{1}{2}$ by $\frac{3}{4}$ of $\frac{3}{5}$ of 6. *Ans.* $\frac{25}{72}$.
10. Divide $\frac{9}{11}$ of $\frac{11}{9}$ by $\frac{7}{8}$ of $\frac{6}{7}$. *Ans.* $1\frac{1}{3}$

Rule of Three in Vulgar Fractions.

State the given terms as the question suggests, as in the Rule of Three; reduce compound and mixed fractions to simple fractions, and to their least terms; consider which of the quantities should be the divisor, or first antecedent, and invert its terms; then take the continual product of the numerators for the numerator, and the continual product of the denominators for the denominator of the fourth proportional required.

Note. In multiplying, terms which are common should be rejected, and such as have a common measure divided by it, using the quotients.

Examples for the Slate.

1. Find a 4th proportional to the following quantities, viz.: for 2, 3, and $\frac{1}{3}$; 3, 4, and $\frac{1}{4}$; 4, 5, and $\frac{1}{5}$.

$2 : 3 :: \frac{1}{3} : \frac{3}{2} \times \frac{1}{3} = \frac{1}{2}$. *Ans.*
$3 : 4 :: \frac{1}{4} : \frac{4}{3} \times \frac{1}{4} = \frac{1}{3}$. *Ans.*
$4 : 5 :: \frac{1}{5} : \frac{5}{4} \times \frac{1}{5} = \frac{1}{4}$. *Ans.*

2. Find a 4th proportional to the following, viz.: $\frac{1}{2}$, $\frac{1}{3}$, and 3; $\frac{1}{3}$, $\frac{1}{4}$, and 4; $\frac{1}{4}$, $\frac{1}{5}$, and 5.

$\frac{1}{2} : \frac{1}{3} :: 3 : \frac{2}{1} \times \frac{1}{3} \times \frac{3}{1} = 2$. *Ans.* 2.
$\frac{1}{3} : \frac{1}{4} :: 4 : \frac{3}{1} \times \frac{1}{4} \times \frac{4}{1} = 3$. *Ans.* 3.
$\frac{1}{4} : \frac{1}{5} :: 5 : \frac{4}{1} \times \frac{1}{5} \times \frac{5}{1} = 4$. *Ans.* 4.

1811	1812	1813	1814	1815	1816	1817	1818	1819	1820
$\frac{1}{1811}$	$\frac{1}{1812}$	$\frac{1}{1813}$	$\frac{1}{1814}$	$\frac{1}{1815}$	$\frac{1}{1816}$	$\frac{1}{1817}$	$\frac{1}{1818}$	$\frac{1}{1819}$	$\frac{1}{1820}$

3. Find a 4th proportional to the following, viz.: $\frac{1}{2}$, $\frac{1}{7}$, and 7; $\frac{1}{5}$, $\frac{1}{9}$, and 9; $\frac{1}{7}$, $\frac{1}{12}$, and 12.

$\frac{1}{2} : \frac{1}{7} :: 7 : \frac{2}{1}\times\frac{1}{7}\times\frac{7}{1}=2.$ *Ans.* 2.

$\frac{1}{5} : \frac{1}{9} :: 9 : \frac{5}{1}\times\frac{1}{9}\times\frac{9}{1}=5.$ *Ans.* 5.

$\frac{1}{7} : \frac{1}{12} :: 12 : \frac{7}{1}\times\frac{1}{12}\times\frac{12}{1}=7.$ *Ans.* 7.

From these examples it appears that any term in the line of the prime series is to any other term as the reciprocal of the latter is to that of the former. These exercises may be continued upon the lines of numbers and reciprocals until the subject is made familiar to the reader. The reason is apparent upon reducing the two fractions to a common denominator, that is, to units of the same value.

4. Find a 4th proportional to the following, viz.:

$\frac{1}{3}$, $\frac{1}{4}$, and $\frac{1}{5}$. *Ans.* $\frac{3}{20}$. $\frac{1}{4}$, $\frac{1}{5}$, and $\frac{1}{6}$. *Ans.* $\frac{2}{15}$.
$\frac{1}{6}$, $\frac{1}{7}$, and $\frac{1}{8}$. *Ans.* $\frac{3}{28}$. $\frac{1}{7}$, $\frac{1}{8}$, and $\frac{1}{9}$. *Ans.* $\frac{7}{72}$.
$\frac{1}{9}$, $\frac{1}{10}$, and $\frac{1}{11}$. *Ans.* $\frac{9}{110}$. $\frac{1}{11}$, $\frac{1}{12}$, and $\frac{1}{13}$. *Ans.* $\frac{11}{156}$.

5. If $\frac{3}{4}$ of a yard of cloth cost $\frac{15}{16}$\$, how much will $\frac{2}{3}$ of a yard of the same cost? *Ans.* \$83$\frac{1}{3}$.

6. If $\frac{3}{8}$ yard of velvet cost $\frac{2}{5}$£, what will $\frac{5}{16}$ yard amount to? *Ans.* 6s. 8d.

7. What will 3$\frac{1}{2}$ oz. of American coined silver be worth, if 1 oz. be valued at 116$\frac{12}{43}$ cents? *Ans.* \$4.06$\frac{4}{43}$.

8. If $\frac{3}{16}$ of a ship be worth \$1400, what sum may the half owner demand for his share? *Ans.* \$3733.33$\frac{1}{3}$.

9. If 1 cwt. cost 15$\frac{3}{4}$s. what will $\frac{1}{4}$ of 1 cwt. amount to? *Ans.* 3s. 11$\frac{1}{4}$d.

10. What will 12$\frac{1}{2}$ gallons of wine come to at \$$\frac{3}{16}$ per pint? *Ans.* \$18.75.

11. What cost 5$\frac{3}{4}$ lbs. of tea at $\frac{3}{8}$ dol. for $\frac{1}{4}$ lb.? *Ans.* 8$\frac{5}{8}$ dols.

12. If $\frac{4}{7}$ oz. avoirdupois cost $\frac{5}{16}$ dol., what will $\frac{5}{14}$ lb. amount to? *Ans.* 3$\frac{1}{8}$ dol.

13. A merchant owning $\frac{3}{4}$ of a ship sells $\frac{1}{4}$ of his part for 250£. Query the whole value of the ship at this rate? *Ans.* 1333£ 6s. 8d.

14. How many stones of 1$\frac{3}{4}$ ft. long, $\frac{2}{3}$ ft. broad, and $\frac{3}{4}$ ft. thick, are equal to 50 stones 3$\frac{1}{2}$ feet long, 2$\frac{1}{2}$ ft. broad, and 1$\frac{1}{5}$ ft. thick? *Ans.* 600.

15. What quantity of shalloon that is $\frac{3}{4}$ yd. wide will line 9$\frac{1}{2}$ yds. cloth that is 2$\frac{1}{2}$ yds. wide? *Ans.* 31$\frac{2}{3}$ yds.

1821	1822	1823	1824	1825	1826	1827	1828	1829	1830
$\frac{1}{1821}$	$\frac{1}{1822}$	$\frac{1}{1823}$	$\frac{1}{1824}$	$\frac{1}{1825}$	$\frac{1}{1826}$	$\frac{1}{1827}$	$\frac{1}{1828}$	$\frac{1}{1829}$	$\frac{1}{1830}$

16. What length of board that is $7\frac{3}{4}$ inches broad will be requisite to make a square foot? *Ans.* $18\frac{13}{31}$ in.

17. The old eagle weighs 11 dwts. 6 grs. of gold, therefore, how many such pieces will 4 lbs. 7 oz. 13 dwts. 18 grs. of gold of the same fineness make? *Ans.* 99.

18. If $6\frac{1}{2}$ yds. cloth that is $\frac{3}{5}$ yd. wide will line a cloak, what quantity that is $\frac{4}{5}$ yd. wide will do the same?
Ans. $4\frac{7}{8}$ yards.

19. How much cloth $1\frac{1}{5}$ yds. wide will equal $9\frac{3}{4}$ yds. of $\frac{4}{5}$ yd. wide? *Ans.* $6\frac{1}{2}$ yds.

20. If 6 men do a job of work in $\frac{3}{8}$ of a day, in what time will 20 men accomplish a similar work?
Ans. 1 hr. 21 m.

21. 1 foot thick of 144 superficial inches makes 1 cubic foot; then how thick must a stone be whose surface measures $\frac{3}{4}$ as much to make the same? *Ans.* $1\frac{1}{3}$ ft.

22. If a plane of 108 sq. inches require a depth of 32 inches to measure 2 cubic ft., how much surface will $\frac{3}{4}$ of the thickness require to measure the same?
Ans. 144 sq. inches.

23. If I lend my friend 168 dols. for $9\frac{3}{4}$ months, what sum should he lend me $6\frac{1}{2}$ months to be equivalent?
Ans. 252 dols.

24. What will 2 cords of wood amount to, if $\frac{1}{4}$ of $\frac{4}{5}$ of $\frac{5}{8}$ of a cord cost $\frac{5}{6}$ of $\frac{9}{10}$ of a dol. *Ans.* 12 dols.

Double Rule of Three in Vulgar Fractions.

RULE. State the given quantities as in Double Rule of Three, page 109; reduce them to simple fractions, invert the divisors, then, rejecting common terms, take the product of the numerators for the numerator, and the product of the denominators for the denominator of the answer.

Or, reduce similar terms to a common denominator; that is, to units of the same value, reject the com. denominator, and use the numerators as multiples instead of parts.

Examples for the Slate.

1. Find a consequent to which 3 shall have the compound ratio of $\frac{1}{2}$ to $\frac{2}{3}$, and $\frac{3}{4}$ to $\frac{4}{5}$?

1831	1832	1833	1834	1835	1836	1837	1838	1839	1840
$\frac{1}{1831}$	$\frac{1}{1832}$	$\frac{1}{1833}$	$\frac{1}{1824}$	$\frac{1}{1835}$	$\frac{1}{1836}$	$\frac{1}{1837}$	$\frac{1}{1838}$	$\frac{1}{1839}$	$\frac{1}{1840}$

$\frac{1}{2}$ and $\frac{3}{4}$ are the antecedents or divisors; invert them; then, $\frac{2}{1}\times\frac{4}{3}\times\frac{2}{3}\times\frac{4}{5}\times\frac{3}{1}=\frac{64}{15}=4\frac{4}{15}$ *Ans.*

Or thus, $\frac{1}{2} : \frac{2}{3} :: \frac{3}{6} : \frac{4}{6}$ or $3 : 4$

$\frac{3}{4} : \frac{4}{5} :: \frac{15}{20} : \frac{16}{20}$ or $15 : 16$

The compound ratio is $45 : 64 :: 3 : 4\frac{4}{15}$. *Ans.*

2. Find a consequent to which 5 shall have the compound ratio of $\frac{9}{10}$ to $\frac{7}{8}$, and $\frac{5}{6}$ to $\frac{6}{7}$. *Ans.* 5.

3. If $1\frac{1}{2}$ yard of $\frac{8}{4}$ yard wide cost \$8, how much should 6 yards of cloth of $\frac{4}{4}$ yard wide cost? *Ans.* \$16.

4. If when coffee is $\frac{1}{8}$\$ per lb., 8 persons spend $1\frac{1}{4}$\$ per month in that article, how much will 6 persons spend per month when the price is \$$\frac{3}{16}$ per lb.? *Ans.* \$1.40$\frac{5}{8}$.

5. If \$$2\frac{2}{5}$ pay for 4 yards of cloth $\frac{3}{4}$ yd. wide, how many yards of $\frac{8}{4}$ wide will \$$9\frac{3}{5}$ purchase? *Ans.* 6 yds.

6. If 6 students spend \$$5\frac{7}{18}$ in 9 days, what sum will 18 students spend in 30 days? *Ans.* \$$53\frac{8}{9}$.

7. The price of 6 loaves of $\frac{1}{2}$ lb. each being 30 cents, how much will 9 loaves of $\frac{3}{4}$ lb. each come to?

Ans. $67\frac{1}{2}$ cents.

8. If \$150 gain \$$4\frac{11}{16}$ in $7\frac{1}{2}$ months, what interest will \$75 gain in $3\frac{3}{4}$ months? *Ans.* \$$1\frac{11}{64}$.

9. If \$150 in $\frac{3}{4}$ of a year gain \$$5\frac{1}{16}$, at what rate per cent. per annum is the interest computed?

Ans. $4\frac{1}{2}$ per cent.

10. If \$100 in one year gain \$$6\frac{3}{4}$, what interest will \$75 produce in $\frac{7}{8}$ of a year? *Ans.* \$$4\frac{55}{128}$.

11. If 12 persons expend \$624 in $\frac{1}{4}$ year, what sum will 5 persons require for $\frac{3}{26}$ of a year? *Ans.* \$120.

12. If 10 men receive \$2700 for $\frac{3}{4}$ year's service, how much is due to 4 men who have served $\frac{3}{8}$ of a year?

Ans. \$540.

13. If 14 yds. of calico $\frac{1}{2}$ yard wide make 2 dresses, how many yards of gingham $\frac{5}{8}$ yd. wide will be required to make 6 dresses? *Ans.* $33\frac{3}{5}$ yds.

14. A P of $3\frac{1}{2}$ lbs. suspended $2\frac{1}{2}$ feet from the centre of motion of a lever equipoises $\frac{3}{4}$ cwt. at the other end, then how much would a P of 6 lbs. draw, being placed $3\frac{3}{4}$ feet from said centre. *Ans.* 1 cwt. 3 qrs. 20 lbs.

1841	1842	1843	1844	1845	1846	1847	1848	1849	1850
$\frac{1}{1841}$	$\frac{1}{1842}$	$\frac{1}{1843}$	$\frac{1}{1844}$	$\frac{1}{1845}$	$\frac{1}{1846}$	$\frac{1}{1847}$	$\frac{1}{1848}$	$\frac{1}{1849}$	$\frac{1}{1850}$

INVOLUTION, OR POWERS.

The powers of a number are expressed by the series of ordinal numbers, viz.: first, second, third, fourth, fifth, sixth, seventh, eighth, ninth, tenth, &c. These terms are represented by the figures 1, 2, 3, 4, &c., which are called indices. In table 1, Compound Interest, following, the indices represent years, and are ranged in column on the left margin of the table, opposite the powers of the several ratios there involved. But the schools usually place the indices or exponents over the root.

Besides, by the numerical index, powers are sometimes designated by names which express in part their relation to each other, viz.:

1. The root,
2. The square,
3. the cube,
4. The biquadrate,
5. The sursolid,
6. The squared cube,
7. The 2d sursolid,
8. The sq'd biquadrate,
9. The cubed cube,
10. The sq'd sursolid;

and so on without any termination.

In involution, the root may be any multiple or part of a unit. In the decimal notation the root is 10, for multiples, and $\frac{1}{10}$ for parts of a unit.

Thus, 10^1, 10^2, 10^3, 10^4, &c., in the line of increase; and $(\frac{1}{10})^1$, $(\frac{1}{10})^2$, $(\frac{1}{10})^3$, $(\frac{1}{10})^4$, &c., in the line of decrease.

The intervening unit cannot be a power of either of these series; but when they become a connected geometrical series, it forms one of the terms, thus:

+4	+3	+2	+1	+0	—1	—2	—3	—4
10000,	1000,	100,	10,	$\overline{1}$,	.1,	.01,	.001,	.0001.

The indices above show the distance of each significant figure from the units' rank. Those on the left, being indices of multiples, are distinguished by the positive sign +; while those on the right, being indices of parts, are marked by the negative sign —. The unit

1851	1852	1853	1854	1855	1856	1857	1858	1859	1860
$\frac{1}{1851}$	$\frac{1}{1852}$	$\frac{1}{1853}$	$\frac{1}{1854}$	$\frac{1}{1855}$	$\frac{1}{1856}$	$\frac{1}{1857}$	$\frac{1}{1858}$	$\frac{1}{1859}$	$\frac{1}{1860}$

claiming relation to both sides, is marked by the double sign + and —.

These indices form the basis of the tables of logarithms invented by Baron Napier of Scotland.

The sum of two indices is the index of the product of their powers: this is in part indicated by the names of the powers. But it is easy to see, for example, that $2^1\times2^1\times2^1\times2^1=2^2\times2^2=2^1\times2^3=2^4=16$.

The double of an index is the index of the square of a power; thus, $2^{2\times2}=2^4=16$; and 3 times the index of a power is the index of the cube of that power; thus, $2^{2\times3}=2^6=2^2\times2^2\times2^2=4\times4\times4=64$.

Class Exercise in forming Powers.

1. Multiply 2 by 2, and the product by 2, and so on to 2048; range the powers in a line, and set over each its proper index.
2. Multiply 3 by 3, and the product by 3, and so on to 2187; range the powers in a line, and set over each its proper index.
3. Form, in like manner, the powers of 4 to 2048, and set over each its proper index.
4. Form, in like manner, the series of the powers of 5, to the 5th power, and annex their indices.
5. Form, in like manner, the powers of 6, to the 4th power, and annex their indices.
6. Form also 4 powers of 7 and annex their indices.
7. Form also 4 powers of 8 and annex their indices.
8. Form also 4 powers of 9 and annex their indices.

Having all these powers and their indices ranged on slates; Query what power of 4 is equal to the 4th of 2? what power of 4 equals the 4th of 8? what power of 2 equals the 2d of 8? what power of 2 equals the 4th of 8? what powers of 3 equal the 2d and 4th of 9? Continue these comparisons until the idea is distinctly formed in the mind respecting the sum of the indices in accordance with the product of the powers.

The involution of ratios is, however, the most important application of this rule. In multiplication of deci-

1861	1862	1863	1864	1865	1866	1867	1868	1869	1870
$\frac{1}{1861}$	$\frac{1}{1862}$	$\frac{1}{1863}$	$\frac{1}{1864}$	$\frac{1}{1865}$	$\frac{1}{1866}$	$\frac{1}{1867}$	$\frac{1}{1868}$	$\frac{1}{1869}$	$\frac{1}{1870}$

mals, pages 42, 43, rules for contracting products are given, and a few questions of involution follow, on page 44. It is now in order to extend the practice there begun, so as to form a table of the amounts of $1, at various rates per cent. for 20 years; the years being the indices of the powers.

Examples for the Slate.

1. What are the several powers of 1.03, up to the 20th?

1.03	1	1.159274	5	1.425761	12
1.03	1	30.1	1	42773	
309		1.159274		1.468534	13
1030		34778		44056	
1.0609	2	1.194052	6	1.512590	14
1.03	1	35822		45378	
31827		1.229874	7	1.557968	15
10609		36896		46739	
1.092727	3	1.266770	8	1.604707	16
30.1	1	38003		48141	
1.092727		1.304773	9	1.652848	17
32782		39143		49586	
1.125509	4	1.343916	10	1.702434	18
30.1	1	40318		51073	
1.125509		1.384234	11	1.753507	19
33765		41527		52605	
1.159274	5	1.425761	12	1.806112	20

Note. In the preceding operations, and after the third power is raised, the multiplier is inverted so as to retain only six places of parts in the products; also after the sixth power is raised the multiplier is omitted, to save the repetition of two lines, viz.: the line of the multiplier and that of the product of 1, which is the same as the multiplicand.

1871	1872	1873	1874	1875	1876	1877	1878	1879	1880
$\frac{1}{1871}$	$\frac{1}{1872}$	$\frac{1}{1873}$	$\frac{1}{1874}$	$\frac{1}{1875}$	$\frac{1}{1876}$	$\frac{1}{1877}$	$\frac{1}{1878}$	$\frac{1}{1879}$	$\frac{1}{1880}$

2. After the same manner, raise 20 powers of 1.035, and mark them with the indices.

3. After the same manner, raise 20 powers of 1.04, and mark them with the indices.

4. After the same manner, raise 20 powers of 1.045, and mark them with the indices.

5. After the same manner, raise 20 powers of 1.05, and mark them with the indices.

6. After the same manner, raise 20 powers of 1.055, and mark them with the indices.

7. After the same manner raise 20 powers of 1.06, and mark them with the indices.

The answers to these operations form the columns of Table 1, Compound Interest. There, however, only five places of parts are retained, for want of space: but it is obvious that the more places we retain the nearer we may approximate to the true result.

Remark. At page 47 and 48, it was required to divide $1 by 1.03, by 1.035, by 1.04, by 1.045, by 1.05, by 1.055 and by 1.06, successively, twenty times, and to reserve the quotients for the purpose of forming a table of the present worths of $1. Now, if the first quotient in each example, that is, the quotient of $1 by 1.03, of $1 by 1.035, &c., were raised to 20 powers by involution, these 20 powers should be the same as the 20 quotients by division.—If, therefore, the class are willing to work, they may find sufficient employment for several hours, in raising 20 powers of the aforesaid first quotients. We do not however prescribe it as a task.

The powers of the numerator and denominator of a fraction are formed separately; unless when the fraction is reduced to a decimal; in which case the denominator of the power is known by the rules of decimal multiplication. See pages 42, 43.

Examples for the Slate.

What is the square of $\frac{1}{2}$?
What is the square of $\frac{2}{3}$?
What is the square of $\frac{3}{4}$?
What is the square of $\frac{4}{5}$?
What is the square of $\frac{5}{6}$?
What is the square of $\frac{6}{7}$?
What is the square of $\frac{7}{8}$?
What is the square of $\frac{8}{9}$?

What is the cube of $\frac{1}{2}$?
What is the cube of $\frac{2}{3}$?
What is the cube of $\frac{3}{4}$?
What is the cube of $\frac{4}{5}$?
What is the cube of $\frac{5}{6}$?
What is the cube of $\frac{6}{7}$?
What is the cube of $\frac{7}{8}$?
What is the cube of $\frac{8}{9}$?

1881	1882	1883	1884	1885	1886	1887	1888	1889	1890
$\frac{1}{1881}$	$\frac{1}{1882}$	$\frac{1}{1883}$	$\frac{1}{1884}$	$\frac{1}{1885}$	$\frac{1}{1886}$	$\frac{1}{1887}$	$\frac{1}{1888}$	$\frac{1}{1889}$	$\frac{1}{1890}$

EVOLUTION; OR, EXTRACTING OF ROOTS.

Rational roots are such as may be exactly found; the roots of imperfect powers are called *surd* roots.

The square of a single figure cannot exceed *two* places; neither can the cube exceed *three* places; for 9 is the greatest single figure, and its square is 81, only *two* places; and its cube 729, only three places.

The same rule applies to higher powers: wherefore, a power, the root of which is to be extracted, must be distinguished into *two* places for the square root, *three* places for the cube root, *four* places for the 4th root, &c. The count begins at units for the multiples, and at tenths for the parts; and in decimal parts, the periods are completed with 0s. The root will have as many places as the power has periods.

To extract the square root is to find the first power of a number, when the second is given; or, in extensions, to find one side of a square.

RULE 1. Distinguish the given power into periods of two places; subtract from the highest period the greatest square it contains, setting the root as a quotient; join the next period to the remainder for a dividual.

2. Take twice the root for a divisor; seek how often the divisor is contained in the dividual, rejecting its units' figure; annex the result to both root and divisor; subtract from the dividual the product of the divisor multiplied by the last quotient figure; bring down the next period to the remainder.

3. The root $\times 2$, or the last divisor $+$ the last figure of the root, will be the defective divisor; a new quotient figure is sought, and the same process is repeated to the end.

Note. After finding 3 or 4 decimal places, proceed as in division of decimals contracted.

1891	1892	1893	1894	1895	1896	1897	1898	1899	1900
$\frac{1}{1891}$	$\frac{1}{1892}$	$\frac{1}{1893}$	$\frac{1}{1894}$	$\frac{1}{1895}$	$\frac{1}{1896}$	$\frac{1}{1897}$	$\frac{1}{1898}$	$\frac{1}{1899}$	$\frac{1}{1900}$

The class respond to the following questions, viz.:

What is the square of 1? *Ans.* A square 1.
What is the square of 2? *Ans.* Four square ones.
—the square of 3?
—the square of 4?
—the square of 5?
—the square of 6?
—the square of 7?
—the square of 8?
—the square of 9?

Examples for the Slate.

1. What is the square root of 2, to 9 places of figures?

$\sqrt{2}$=(1.41421356 *Ans.*

	1^2=1	
2×1 and 4 annexed.	24 | 100 4 | 96	100, 400, 11900, 60400 are the remainders, with a period of two ciphers annexed to each, for want of figures in the power to be brought down.
2×14 and 1 annexed.	281 | 400 1 | 281	
2×141 and 4 annexed.	2824 | 11900 4 | 11296	
2×1414, and 2 annexed.	28282 | 60400 2 | 56564	At 3836 the contraction begins; where figures are dropped from the divisor.
2×root is the defective divisor, the figure annexed is the result of the trial, how oft? The 4, 1, 4, 2, added to the former divisors, make the following defective divisors respectively; the same as doubling the root.	28284) 3836(1356 .... 2828 1008 848 160 141 19 17	

1901	1902	1903	1904	1905	1906	1907	1908	1909	1910
$\frac{1}{1901}$	$\frac{1}{1902}$	$\frac{1}{1903}$	$\frac{1}{1904}$	$\frac{1}{1905}$	$\frac{1}{1906}$	$\frac{1}{1907}$	$\frac{1}{1908}$	$\frac{1}{1909}$	$\frac{1}{1910}$

Leading Questions in the work of the annexed examples.

What is the greatest square contained in 20? in 7? in 11? in 4? in 90? in 92? in 99? in 13? in 2? in 8? in 3? in 14? in 4? in 22? in 5? in 32? in 7? in 59?

2. What is the square root of 746496?

```
√746496 = 864. Ans.
 64
 --
166)1064
     996
    ----
1724) 6896
      6896
      ====
```

3. What is the square root of 2025? *Ans.* 45.
4. What is the square root of .000729? *Ans.* .027.
5. What is the square root of 11? *Ans.* 3.316624.
6. What is the square root of 40401? *Ans.* 201.
7. What is the square root of 900601? *Ans.* 949.
8. What is the square root of 929296? *Ans.* 964.
9. What is the square root of 998001? *Ans.* 999.
10. What is the square root of 138384? *Ans.* 372.
11. What is the square root of 287? *Ans.* 16.9410743.
12. What is the square root of 82369? *Ans.* 287.
13. What is the square root of 384? *Ans.* 19.5959179.
14. What is the square root of 147456? *Ans.* 384.
15. What is the square root of 476? *Ans.* 21.8174242.
16 What is the square root of 226576? *Ans.* 476.
17. What is the square root of 572? *Ans.* 23.9165215
18. What is the square root of 327184? *Ans.* 572.
19. What is the square root of 774? *Ans.* 27.8208555.
20. What is the square root of 599076? *Ans.* 774.

To extract the Square Root of a Vulgar Fraction.

RULE. Reduce the fraction to its lowest terms, and take the root of the numerator for the numerator, and of the denominator for the denominator of the fractional root

1911	1912	1913	1914	1915	1916	1917	1918	1919	1920
$\frac{1}{1911}$	$\frac{1}{1912}$	$\frac{1}{1913}$	$\frac{1}{1914}$	$\frac{1}{1915}$	$\frac{1}{1916}$	$\frac{1}{1917}$	$\frac{1}{1918}$	$\frac{1}{1919}$	$\frac{1}{1920}$

required; but if the fraction be a surd power, reduce it to a decimal one, and extract the root.

Reduce a mixed number to an improper fraction, and a compound fraction to a simple one, and either of them to the decimal form, if surds; then extract the root.

Questions requiring Oral Answers.

What is the sq. root of $\frac{1}{4}$? What is the sq. root of $\frac{25}{36}$?
What is the sq. root of $\frac{4}{9}$? What is the sq. root of $\frac{36}{49}$?
What is the sq. root of $\frac{9}{16}$? What is the sq. root of $\frac{49}{64}$?
What is the sq. root of $\frac{16}{25}$? What is the sq. root of $\frac{64}{81}$?

Examples for the Slate.

1. What is the square root of $\frac{25}{36}$? *Ans.* $\frac{5}{6}$.
2. What is the square root of $\frac{49}{121}$? *Ans.* $\frac{7}{11}$.
3. What is the square root of $\frac{64}{81}$? *Ans.* $\frac{8}{9}$.
4. What is the square root of $\frac{169}{576}$? *Ans.* $\frac{13}{24}$.
5. What is the square root of $\frac{961}{11881}$? *Ans.* $\frac{31}{109}$.
6. What is the square root of $4\frac{26749}{81225}$? *Ans.* $2\frac{23}{285}$.
7. What is the square root of $3\frac{109}{225}$? *Ans.* 1.8666+.
8. What is the square root of $2\frac{401}{484}$? *Ans.* 1.68+.
9. What is the square root of $12\frac{1}{4}$? *Ans.* 3.5.

Properties of Numbers relating to the Square, Right Angle, Circle, Sphere and Cylinder.

Note 1. The line AC is that which lies between the points A and C; the line AB lies between the points A and B; and the line BC lies between the points B and C.

C

A B

RULE 1. $AC^2=AB^2+BC^2$: therefore when AC and AB are given to find BC; $AC^2-AB^2=BC^2$, and $\sqrt{BC^2}=BC$.

2. When AC and BC are given to find AB; $AC^2-BC^2=AB^2$, and $\sqrt{AB^2}=AB$.

3. When AB and BC are given to find AC; $AB^2+BC^2=AC^2$, and $\sqrt{AC^2}=AC$.

4. The sum of two numbers multiplied by their difference equals the difference of their squares, viz.: $(AC+BC)\times(AC-BC)=AC^2-BC^2$; therefore since $AB^2=AC^2-BC^2$, $AB^2\div(AC+BC)=AC-BC$.

5. Also of two numbers, the greater equals half their sum + half

1921	1922	1923	1924	1925	1926	1927	1928	1929	1930
$\frac{1}{1921}$	$\frac{1}{1922}$	$\frac{1}{1923}$	$\frac{1}{1924}$	$\frac{1}{1925}$	$\frac{1}{1926}$	$\frac{1}{1927}$	$\frac{1}{1928}$	$\frac{1}{1929}$	$\frac{1}{1930}$

their difference, and the less equals half their sum — half their difference: therefore $\frac{1}{2}$ (AC+BC)+$\frac{1}{2}$ (AC—BC)=AC; and $\frac{1}{2}$ (AC+BC) —$\frac{1}{2}$ (AC—BC)=BC.

Note. 2. The area of the greater circle AB, is to the area of the less circle ab, as the square of AB is to the square of ab, that is as the squares of their diameters. Also, the ratios of the circumference of a circle to its diameter, from successive discoveries, are as follows, viz.: 3 to 1, 22 to 7, 333 to 106, 355 to 113, and 314159 to 100000; these are alternately above and below the true ratio.

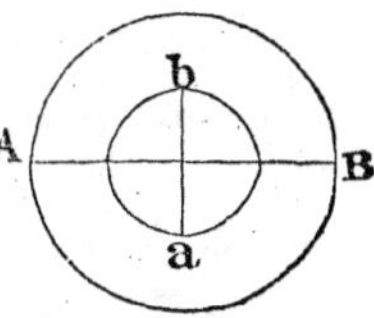

The ratio of 3 to 1 is above the true ratio; but the cooper uses it: he applies his pole 3 times across the head of his vessel, notches it and forms a hoop; he finds this hoop may be driven to its station, and cannot be persuaded of the want of accuracy in the measure.

The ratio of 3.1416 to 1, is the ratio generally used in science, the last two figures 59 being changed to 60, and 0 dropped.

Note 3. In any regular polygon, $\frac{1}{2}$ the perimeter multiplied by the straight line drawn from the centre perpendicular upon one of the sides equals the area: in like manner, $\frac{1}{2}$ the circumference multiplied by the radius equals the area of a circle. And when the diameter of a circle is 1, its circumference will be 3.1416; therefore $\frac{1}{2}\times 1.5708=.7854$; this equals the area of a circle whose diameter is 1.

Again, since the base of a cylinder is a circle, the area of the base $.7854\times 1=$ the solid contents of a cylinder whose height and diameter are 1; also, since a sphere or globe is $\frac{2}{3}$ of its circumscribing cylinder, $.7854\times\frac{2}{3}=.5236$, which is the solid contents of a globe whose diameter is 1. Therefore, also, a globe is to the cube of its diameter as .5236 is to 1^3, which is very nearly as 11 is to 21.

Examples for the Slate.

1. In the triangle ABC, let AC=5 yards, and AB=4 yards; then what is the length of BC?

$\sqrt{(5^2-4^2)}=\sqrt{(25-16)}=\sqrt{9}=3$. *Ans.*

2. Let AC=10 feet, and AB=8 feet, to find BC.

$\sqrt{(10^2-8^2)}=\sqrt{(100-64)}=\sqrt{36}=6$. *Ans.*

3. Let AB=9, BC=12, to find AC.

$\sqrt{(9^2+12^2)}=\sqrt{(81+144)}=\sqrt{225}=15$. *Ans.*

4. Let BC=1½, AB=2, to find AC. *Ans.* 2½.

5. Let AC=20, BC=16, to find AB. *Ans.* 12.

6. Let BC=23 feet 10 inches, and AB=11 feet, how much is the length of AC? *Ans.* 26¼ feet.

1931	1932	1933	1934	1935	1936	1937	1938	1939	1940
$\frac{1}{1931}$	$\frac{1}{1932}$	$\frac{1}{1933}$	$\frac{1}{1934}$	$\frac{1}{1935}$	$\frac{1}{1936}$	$\frac{1}{1937}$	$\frac{1}{1938}$	$\frac{1}{1939}$	$\frac{1}{1940}$

7. Let AC+BC=80, and AB=40, what is the length of AC alone? also, of BC alone? (Rules 4 and 5.)
Ans. AC 50, BC 30.

Note. Carpenters use the 10 feet rod in squaring frames; measuring 8 ft. and 6 ft. on the lines which they intend to form the right angle; because 10 feet is the length of the line which will join the extremities of the 6 feet and 8 feet lines, when the angle is 90° or ¼ circle.

8. The sum of 16s. 4d. was given in pence to a certain number of indigent persons, and to each as many pennies as there were persons; what was the number?
Ans. 14 persons.

9. How many square yards and how many square feet are in one side of a square perch or rod?
Ans. 5½ sq. yards, 16½ sq. ft.

Although a line, long measure, is said to have no breadth, yet the units which form the side of a square are square units, and similar to the whole square.

10. A square mile contains 102400 square poles, then how many square rods does one side of it measure? how many acres does the whole measure?
Ans. 320 rods, 640 acres.

11. A shore set 11 feet from the base of a jamb reaches 23 feet 10 inches of its perpendicular height; what is the length of the shore? *Ans.* 26¼ feet.

12. During a storm a pole set in the ground is broken 39 feet from the top; the broken part adheres, while the top meets the ground 15 feet from the stump, forming a triangle, as ABC; what was the height of the pole before its fall? *Ans.* 75 feet.

13. A circular pipe of 1½ inches diameter will fill a cistern in 5 hours; required the diameter of a pipe that will fill the same in 55.1 minutes? *Ans.* 3.5 inches.

14. A ladder 40 feet long, planted in a street, reaches a window on one side, 33 feet from the ground, and being turned over, reaches the opposite wall at the height of 21 feet; the width of the street is required?
Ans. 56.649 feet.

15. The height of a steeple may be determined by means of a paper kite, for example, while the kite is at

1941	1942	1943	1944	1945	1946	1947	1948	1949	1950
$\frac{1}{1941}$	$\frac{1}{1942}$	$\frac{1}{1943}$	$\frac{1}{1944}$	$\frac{1}{1945}$	$\frac{1}{1946}$	$\frac{1}{1947}$	$\frac{1}{1948}$	$\frac{1}{1949}$	$\frac{1}{1950}$

the top, let the expended line be 250 feet, and the base of the steeple 200 feet, from the boy's station: the perpendicular height is required? *Ans.* 150 feet.

16. "A horse in the midst of a meadow suppose
Made fast to a stake by a line from his nose;
How long should the line be, that, reaching around,
Permits him to graze on an acre of ground?"
Ans. 117 ft. 8 in.

17. The height of the wall that defended a town,
To be 100 feet was by measurement found;
By the side of the wall was erected upright,
A ladder that equaled the fortress in height;
A wag for amusement approaching the place,
Drew the foot of the ladder 10 feet from the base;
Then how far did the *top* of the ladder descend,
Is the question proposed to my juvenile friend?
Ans. 6.015+inches.

THE CUBE ROOT.

A cube has 3 equal dimensions, length, breadth and thickness: to extract the cube root is to find one of these dimensions, having the continual product of the three given. See Evolution.

Rule. Distinguish the given cube into periods of three places; subtract from the leading period the greatest cube it contains; set the root as a quotient, and join the next period to the remainder for a dividual.

Take the root$^2\times 3$ for a divisor; seek how often it is contained in the dividual, rejecting units and tens; annex the result to the root, and its square to the right of the divisor, with a 0 in the tens' place when the square is less than 10.

Add to the divisor the product of the last figure of the root $\times$ the rest $\times 30$; multiply the sum by the said last figure, and subtract the product from the dividual.

Bring down the next period to the remainder; or join as many 0s as will complete the last period of decimals, or a whole period of ciphers, 000, to extend the root:

1951	1952	1953	1954	1955	1956	1957	1958	1959	1960
$\frac{1}{1951}$	$\frac{1}{1952}$	$\frac{1}{1953}$	$\frac{1}{1954}$	$\frac{1}{1955}$	$\frac{1}{1956}$	$\frac{1}{1957}$	$\frac{1}{1958}$	$\frac{1}{1959}$	$\frac{1}{1960}$

find a new divisor, and continue the same process to the end.

Leading Questions in the work of the Examples below.

What is the greatest cube contained in 275? in 68? in 941? in 7? in 15? in 41? in 63? in 124? in 206? in 317? in 682? in 591? in 564? in 480? in 252? in 150? in 100? in 10?

And at the time of giving the oral answers to these questions, announce also the root of the greatest cube.

Examples for the Slate.

1. What is the cube root of 275894.451?

$$\sqrt[3]{275894.451}=65.1$$

	$6^3=216$	
$6^2\times3$ and $5^2=$	10825	59894
$5\times6\times30=$	900	
	11725	58625
$65^2\times3$ and $1^2=$	1267501	1269451
$1\times65\times30=$	1950	
	1269451	1269451

The first period contains $216=6^3$. The first divisor is $6^2\times3=108$; the figure 5 is the result of the trial how oft 108 is contained in 598; then $5^2=25$ is annexed to 108; the divisor is increased by the product of $5\times6\times30=900$; the complete divisor is 11725, which is $\times5$.

2. What is the cube root of 68921? *Ans.* 41.
3. What is the cube root of 941192? *Ans.* 98.
4. What is the cube root of 7880599? *Ans.* 199.
5. What is the cube root of 15252992? *Ans.* 248.
6. What is the cube root of 41421736? *Ans.* 346.
7. What is the cube root of 63044792? *Ans.* 398.
8. What is the cube root of 124251499? *Ans.* 499.
9. What is the cube root of 206425071? *Ans.* 591.
10. What is the cube root of 317214568? *Ans.* 68.2.
11. What is the cube root of 682? *Ans.* 8.802272.
12. What is the cube root of 591? *Ans.* 8.391942.

1961	1962	1963	1964	1965	1966	1967	1968	1969	1970
$\frac{1}{1961}$	$\frac{1}{1962}$	$\frac{1}{1963}$	$\frac{1}{1964}$	$\frac{1}{1965}$	$\frac{1}{1966}$	$\frac{1}{1967}$	$\frac{1}{1968}$	$\frac{1}{1969}$	$\frac{1}{1970}$

13. What is the cube root of 564? *Ans.* 8.262149.
14. What is the cube root of 480? *Ans.* 7.829735.
15. What is the cube root of 252? *Ans.* 6.316359.
16. What is the cube root of 150? *Ans.* 5.313293.
17. What is the cube root of 100? *Ans.* 4.641589.
18. What is the cube root of 10? *Ans.* 2.154435.

To extract the cube root of a vulgar fraction, observe the rule on page 197, which is general.

Examples for the Slate.

1. What is the cube root of $\frac{1}{8}$? of $\frac{8}{27}$? of $\frac{27}{64}$?
   Ans. $\sqrt[3]{\frac{1}{8}}=\frac{1}{2}$. *Ans.* $\sqrt[3]{\frac{8}{27}}=\frac{2}{3}$. *Ans.* $\sqrt[3]{\frac{27}{64}}=\frac{3}{4}$.
2. What is the cube root of $\frac{64}{125}$? of $\frac{125}{216}$? of $\frac{216}{343}$?
3. What is the cube root of $\frac{343}{512}$? of $\frac{512}{729}$? of $\frac{729}{1000}$?
4. What is the cube root of $\frac{686}{1458}$? *Ans.* $\frac{7}{9}$.
5. What is the cube root of $\frac{26730899}{6331625}$? *Ans.* $1\frac{114}{165}$.
6. What is the cube root of $14\frac{1310}{6859}$? *Ans.* $2\frac{8}{19}$.
7. What is the cube root of $92\frac{4303}{24389}$? *Ans.* $4\frac{15}{29}$.
8. What is the cube root of $7\frac{1}{5}$? *Ans.* 1.93+.
9. What is the cube root of $8\frac{5}{7}$? *Ans.* 2.057+.
10. What is the cube root of $\frac{5}{9}$? *Ans.* .82207.
11. What is the cube root of $56623\frac{13}{125}$? *Ans.* 38.4.
12. What is the cube root of $5333\frac{1}{3}$ *Ans.* 17.471.
13. How many cubes of 6 inches are in 1 cubic foot? How many of 4 inches? How many of 3 inches?
    Ans. 8 cubes of 6 in., 27 of 4 in., 64 of 3 in.
14. How much length of iron wire of $\frac{1}{4}$ inch square will one cubic foot of iron make? *Ans.* 2304 feet.
15. A cube contains 970299 cubic inches; how many square inches are in one side of this cube?
    Ans. 9801 sq. in.
16. A log of mahogany contains 110592 cubic inches; what are its dimensions? *Ans.* 4 feet each way.
17. The solidity of a globe is to the cube of its diameter as 11 is to 21; then what is the solid content of a globe whose diameter is 18 inches?
    Ans. $3054\frac{6}{7}$ cubic inches.
18. A cannon shot contains $268\frac{4}{21}$, cubic inches; wh is its diameter? *Ans.* 8 inches.

1971	1972	1973	1974	1975	1976	1977	1978	1979	1980
$\frac{1}{1971}$	$\frac{1}{1972}$	$\frac{1}{1973}$	$\frac{1}{1974}$	$\frac{1}{1975}$	$\frac{1}{1976}$	$\frac{1}{1977}$	$\frac{1}{1978}$	$\frac{1}{1979}$	$\frac{1}{1980}$

19. The diam.^{3} of a globe is to its solidity as 1^3 is to .5236; then what is the content of the earth in cubic miles, whose diameter is 7970 miles long?

Ans. 265078559622.8 miles.

20. When the diameter of a cylinder is 36 inches,—and height 4 inches, what is the solid content?

Ans. $2\frac{1}{4}$ cu. ft.

A General Rule for extracting the Roots of all Powers.

1. Having pointed the power into periods, as the index directs; subtract from the first, the greatest power it contains of the kind, and join the first figure of the next period to the remainder for a dividend.
2. Take the root involved to the given power less one, and this×index for a divisor; find a quotient figure, and annex it to the root.
3. Involve the present root into the given power for a subtrahend, which subtract from the periods of the given power now engaged in the operation.
4. Join the first figure of the next period to the remainder; find a divisor as before, and repeat the same process to the end.

This rule is clearly illustrated by the following

Examples for the Slate.

1. What is the 4th root of 34828517376?

$$34828517376(432 \text{ Ans.}$$

$4^4 =$ 256

Root Index.

$4^3 \times 4 =$ divisor $=$ 256) 922 dividend.

$43^4 =$ 3418801 subtrahend.

Root Index.

$43^3 \times 4 =$ divisor $=$ 318028) 640507 dividend.

$432^4 =$ 34828517376 subtrahend.

1981	1982	1983	1984	1985	1986	1987	1988	1989	1990
$\frac{1}{1981}$	$\frac{1}{1982}$	$\frac{1}{1983}$	$\frac{1}{1984}$	$\frac{1}{1985}$	$\frac{1}{1986}$	$\frac{1}{1987}$	$\frac{1}{1988}$	$\frac{1}{1989}$	$\frac{1}{1990}$

2. What is the 5th root of 380204032? *Ans.* 52.
3. What is the 6th root of 21035.8? *Ans.* 5.254037.
4. What is the 7th root of 21035.8? *Ans.* 4.145392.
5. What is the 8th root of 21035.8? *Ans.* 3.470328.
6. What is the 9th root of 140608? *Ans.* 3.7325+.

Note. When the index of a power is an even number, to extract the square root thereof will depress it to a power half as high: and when the index of a power is divisible by 3, to extract its cube root will depress it to a power $\frac{1}{3}$ as high.

ARITHMETICAL PROGRESSION.

The class of series called Arithmetical consist of numbers which increase or decrease by a common difference.

Their operations depend upon the extreme and mean terms, number of terms, common difference, and sum of the series.

Let a be the least term, g the greatest term,
n number of terms, d common difference,
and s the sum of all the terms.

The terms a and g are the extremes, the intervening terms are called the means.

The sum of the extremes is equal to that of any two means equidistant from them; or, in an odd number of terms, to twice the middle one.

Let 2, 3, 4, 5, 6, be a series of 5 terms, odd;
and 6, 5, 4, 3, 2, the same inverted.

Then, 6+2, the extremes, =5+3, equidistant means, =4+4, twice the middle term.

The prime series is arithmetical, and so are any terms of that series which are equidistant from one another.

1991	1992	1993	1994	1995	1996	1997	1998	1999	2000
$\frac{1}{1991}$	$\frac{1}{1992}$	$\frac{1}{1993}$	$\frac{1}{1994}$	$\frac{1}{1995}$	$\frac{1}{1996}$	$\frac{1}{1997}$	$\frac{1}{1998}$	$\frac{1}{1999}$	$\frac{1}{2000}$

Case 1. When the extremes and number of terms are given to find the sum of the series.

RULE. Multiply the sum of the extremes by the number of terms, and divide by 2; because the product is the sum of two series.

Formula. $(a+g)\times\frac{1}{2}n=s$, or $\frac{1}{2}(a+g)\times n=s$.

Examples for the Slate.

1. The extreme terms of a series are 3 and 19, the number of terms 9; what is the sum of the series?

$\frac{1}{2}(19+3)\times9=11\times9=99$. *Ans.*

2. The extremes are 3 and 36, and the number of terms 12; what is the sum of the series? *Ans.* 234.

3. The extremes are 4 and 56, the number of terms 14; what is the sum of the series? *Ans.* 420.

4. The extremes are 2 and 62; the number of terms 13; what is the sum of the series? *Ans.* 416.

5. The extremes are $\frac{1}{2}$ and 3, the number of terms 16; what is the sum of the series? *Ans.* 28.

6. The extremes are 2 and $6\frac{3}{4}$, the number of terms 20; what is the sum of the series? *Ans.* $87\frac{1}{2}$.

7. The extremes are 3 and 10, the number of terms 22; what is the sum of the series? *Ans.* 143.

8. How many strokes does a regular clock strike in 12 hours? in 24 hours? in 7 days?

9. The clocks of Venice make but one series of strokes in 24 hours; how many strokes do they strike in 30 days more than they would if they renewed the series every 12 hours? *Ans.* 4320.

10. What is the sum of 52 weekly payments, the first being 1s., the last 5£ 3s.? *Ans.* 135.2£.

Case 2. When the extremes and number of terms are given, to find the common difference.

RULE. Find the difference of the extremes, and divide it by the number of terms less 1; because the number of differences is 1 less than the number of terms.

Formula. $(g-a)\div(n-1)=d$.

2001	2002	2003	2004	2005	2006	2007	2008	2009	2010
$\frac{1}{2001}$	$\frac{1}{2002}$	$\frac{1}{2003}$	$\frac{1}{2004}$	$\frac{1}{2005}$	$\frac{1}{2006}$	$\frac{1}{2007}$	$\frac{1}{2008}$	$\frac{1}{2009}$	$\frac{1}{2010}$

Examples for the Slate.

1. The extremes being 3 and 19, and the number of terms 9; what is the common difference?

$$\frac{19-3}{9-1}=\frac{16}{8}=2. \textit{ Ans.}$$

2. The extremes are 3 and 36, and the number of terms 12; what is the common difference? *Ans.* 3.

3. The extremes are 4 and 56, and the number of terms 14; what is the common difference? *Ans.* 4.

What is the common difference in the annexed cases?

4. When a is 2, g 62, and n 13? *Ans.* 5.
5. When a is $\frac{1}{2}$, g 3, and n 16? *Ans.* $\frac{1}{6}$.
6. When a is 2, g $6\frac{3}{4}$, and n 20? *Ans.* $\frac{1}{4}$.
7. When a is 3, g 10, and n 22? *Ans.* $\frac{1}{3}$.
8. When a is 1, g 12, and n 12? *Ans.* 1.
9. When a is 1, g 24, and n 24? *Ans.* 1.
10. When a is 1, g 103, and n 52? *Ans.* 2.

Case 3. When one of the extremes, number of terms, and common difference are given, to find the other extreme.

RULE. Multiply the common difference by the number of terms less 1, the product will be the difference of the extremes, which add to the less, or subtract from the greater; the result will be the greater or less respectively.

Formula. $a+d\times(n-1)=g$; or, $g-d\times(n-1)=a$.

Examples for the Slate.

1. Let a series have 9 terms, of which the least is 3, and the common difference 2; what is the greatest, and the sum of the series?

```
         2 com. dif.
9—1=     8
        ——
        16 dif. extremes.
      + 3 least term.
        ——
Ans.    19 greatest term.
```

Then $\frac{19+3}{2}\times 9=99$ sum.

2011	2012	2013	2014	2015	2016	2017	2018	2019	2020
$\frac{1}{2011}$	$\frac{1}{2012}$	$\frac{1}{2013}$	$\frac{1}{2014}$	$\frac{1}{2015}$	$\frac{1}{2016}$	$\frac{1}{2017}$	$\frac{1}{2018}$	$\frac{1}{2019}$	$\frac{1}{2020}$

2. If the greatest term be 70, common difference 3, and number of terms 21, what is the least term, and sum of the series? *Ans.* a 10, s 840.

3. If the least term be 6, common difference 4, and number of terms 14; what is the greatest term, and sum of all the terms? *Ans.* g 56, s 420.

4. Let a be 2, d 5, and n 13, to find g and s? *Ans.* g 62, s 416.

5. Let g be 3, d $\frac{1}{6}$, and n 16, to find a and s? *Ans.* a $\frac{1}{2}$, s 28.

6. Let a be 2, d $\frac{1}{4}$, and n 20, to find g and s? *Ans.* g $6\frac{3}{4}$, s $87\frac{1}{2}$.

7. Let g be 10, d $\frac{1}{3}$, and n 22, to find a and s? *Ans.* a 3, s 143.

8. Let a be 1, d 1, and n 12, to find g and s? *Ans.* g 12, s 78.

9. Let g be 24, d 1, and n 24, to find a and s? *Ans.* a 1, s 300.

10. Let a be 1, d 2, and n 52, to find g and s? *Ans.* g 103, s 2704.

Case 4. To find any number of means arithmetical between two extremes.

RULE. Since the difference of the extremes is the sum of all the differences, divide it by 2 for *one* mean, by 3 for *two* means, by 4 for *three* means, &c.; the quotient will be the common difference of the terms; which add to the least, or take from the greatest continually: every step in this addition or subtraction reaches one of the required means.

1. Between 2 and 62 what is the single mean?
   62—2= dif. extr.=60; which, divided by 2,=30.
   Then 2+30, or 62—30,=32 the single mean.

2. Between 3 and $\frac{1}{2}$ find two means. *Ans.* $1\frac{1}{3}$, $2\frac{1}{6}$.

3. Between 2 and $6\frac{3}{4}$ find three means. *Ans.* $3\frac{3}{16}$, $4\frac{3}{8}$, $5\frac{9}{16}$.

4. Between 10 and 3 find four means. *Ans.* $4\frac{2}{5}$, $5\frac{4}{5}$, $7\frac{1}{5}$ $8\frac{3}{5}$.

2021	2022	2023	2024	2025	2026	2027	2028	2029	2030
$\frac{1}{2021}$	$\frac{1}{2022}$	$\frac{1}{2023}$	$\frac{1}{2024}$	$\frac{1}{2025}$	$\frac{1}{2026}$	$\frac{1}{2027}$	$\frac{1}{2028}$	$\frac{1}{2029}$	$\frac{1}{2030}$

5. Between 1 and 12 find five means.
   Ans. $2\frac{5}{6}$, $4\frac{2}{3}$, $6\frac{1}{2}$, $8\frac{1}{3}$, $10\frac{1}{6}$.
6. Between 24 and 1 find six means.
   Ans. $4\frac{2}{7}$, $7\frac{4}{7}$, $10\frac{6}{7}$, $14\frac{1}{7}$, $17\frac{3}{7}$, $20\frac{5}{7}$.
7. Between 1 and 52 find seven means.
   Ans. $7\frac{3}{8}$, $13\frac{6}{8}$, $20\frac{1}{8}$, $26\frac{4}{8}$, $32\frac{7}{8}$, $39\frac{2}{8}$, $45\frac{5}{8}$.

Harmonical Proportion exists between 3 or 4 quantities, when the first is to the last as the difference between the first two is to the difference between the last two.

Examples.

1. In the rank 12, 8, 6; 12 : 6 : : 12—8 : 8—6.
2. In the rank 20, 16, 12, 10; 20 : 10 : : 20—16 : 12—10.

GEOMETRICAL PROGRESSION.

The series called Geometrical differs from Involution only in regard to the first term. Here the first term and ratio may be different; there the root is both first term and ratio, or multiplier.

Any number of terms selected from the prime series, or from a series of fractions, which increase or decrease by a constant multiplier, or ratio, form a series in geometrical progression.

The ratio of the series $\frac{1}{8}$, $\frac{1}{4}$, $\frac{1}{2}$, 1, 2, 4, 8, is 2:

The ratio of the series 8, 4, 2, 1, $\frac{1}{2}$, $\frac{1}{4}$, $\frac{1}{8}$, is $\frac{1}{2}$.

The operations belonging to this series depend upon the extreme and mean terms, number of terms, ratio, and the sum of all the terms, or sum of the series.

Let a be the least term, g the greatest term, n number of terms, r the ratio, and s the sum of the series. The extremes are a and g, the intervening terms are called the *means*; and it is plain, from the two series above, that the product of the extremes is equal to the product

2031	2032	2033	2034	2035	2036	2037	2038	2039	2040
$\frac{1}{2031}$	$\frac{1}{2032}$	$\frac{1}{2033}$	$\frac{1}{2034}$	$\frac{1}{2035}$	$\frac{1}{2036}$	$\frac{1}{2037}$	$\frac{1}{2038}$	$\frac{1}{2039}$	$\frac{1}{2040}$

of any two means equidistant from them, or to the square of the middle term, when the number is odd:

For $8\times\frac{1}{8}=4\times\frac{1}{4}=2\times\frac{1}{2}=1\times1$, or 1^2.

Case 1. When one of the extremes, the ratio, and number of terms are given, to find the other extreme.

Rule. Involve the ratio to that power whose index is the number of terms less 1, and multiply the result by the given extreme; the product will be the required term.

Formula. $r^{n-1}\times a=g$. $r^{n-1}\times g=a$.

Examples for the Slate.

1. The least term of a series is $\frac{1}{8}$, ratio 2, number of terms 7; what is the greatest term?

$2^{7-1}\times\frac{1}{8}=2^6\times\frac{1}{8}=64\times\frac{1}{8}=\frac{64}{8}=8$. *Ans.*

2. The greatest term of a series is 8, the number of terms 7, ratio $\frac{1}{2}$; what is the least term?

$(\frac{1}{2})^{7-1}\times8=(\frac{1}{2})^6\times8=\frac{1}{64}\times8=\frac{8}{64}=\frac{1}{8}$ *Ans.*

3. The least term of a series is 1 mill, number of terms 5, ratio 10; what is the greatest term reduced to eagles? *Ans.* 1 eagle.

4. The greatest term of a series is 1 eagle reduced to mills, number of terms 5, ratio $\frac{1}{10}$; what is the least term in mills? *Ans.* 1 mill.

5. The least term of a series is $\frac{1}{27}$, ratio 3, and number of terms 12; what is the greatest term? *Ans.* 6561.

6. The greatest term of a series is 6561, ratio $\frac{1}{3}$, number of terms 12; what is the least term? *Ans.* $\frac{1}{27}$.

7. The greatest term of a series is $\frac{6561}{256}$, the ratio $\frac{2}{3}$, the number of terms 9; what is the least term? *Ans.* 1.

8. The least term of a series is 1, number of terms 9, ratio $1\frac{1}{2}$; what is the greatest term? *Ans.* $25\frac{161}{256}$.

9. A countryman brought to market 1 dozen of fowls; for which he was willing to take what the last would come to; reckoning for the first, 1 cent; for the second, 2 cents; for the third, 4 cents, &c.; what sum should he receive at this rate? *Ans.* $20.48.

2041	2042	2043	2044	2045	2046	2047	2048	2049	2050
$\frac{1}{2041}$	$\frac{1}{2042}$	$\frac{1}{2043}$	$\frac{1}{2044}$	$\frac{1}{2045}$	$\frac{1}{2046}$	$\frac{1}{2047}$	$\frac{1}{2048}$	$\frac{1}{2049}$	$\frac{1}{2050}$

10. Eight persons divide a sum of money among them; the first takes $10, the second $20, the third $40, &c.; how much had the last? *Ans.* $1280.

Case 2. To find the sum of any series in Geometrical Progression, having the extremes, number of terms, and ratio given.

RULE 1. To the greatest term add the difference of the extremes divided by the ratio less 1. Or,

2. To the least term add the difference of the extremes multiplied by the inverted complement* of the ratio.

Formula 1. $g+\overline{(g-a)\div(r-1)}=s$; from Rule 1.

2. $a+\overline{(g-a)\times(\frac{1}{1-r})}=s$; from Rule 2.

Examples for the Slate.

1. The least term of a series is 1, the greatest 64, the ratio 2; what is the sum of the series?

$$64+\frac{64-1}{2-1}=64+\frac{63}{1}=127;\text{ by Rule 1.}$$

Here the series is ascending, because the ratio exceeds 1.

$$1+(64-1)\times\tfrac{2}{1}=1+(63\times2)=127;\text{ by Rule 2.}$$

Here the series is descending, because the ratio is $\frac{1}{2}$.

2. The extremes are $\frac{1}{8}$ and 8, the ratio 2; what is the sum of the series?

$$8+\frac{8-\frac{1}{8}}{2-1}=8+\frac{7\frac{7}{8}}{1}=8+7\tfrac{7}{8}=15\tfrac{7}{8};\text{ by Rule 1.}$$

$$\tfrac{1}{8}+(8-\tfrac{1}{8})\times\tfrac{2}{1}=\tfrac{1}{8}+7\tfrac{7}{8}\times2=\tfrac{1}{8}+15\tfrac{6}{8}=15\tfrac{7}{8};\text{ by Rule 2.}$$

3. The extremes of a series are 1 mill and 1 eagle, the ratio is 10; what is the sum of the series?

---

* The complement to a unit is understood, and the series in the line of decrease, since the ratio is less than *one;* then, if the ratio be $\frac{1}{2}$, the complement$=\frac{1}{2}$; if $\frac{1}{3}$, the comp.$=\frac{2}{3}$; if $\frac{1}{4}$, the comp.$=\frac{3}{4}$, &c. And these complements inverted make $\frac{2}{1}$, $\frac{3}{2}$, $\frac{4}{3}$, &c., but to multiply by a fraction inverted is division.

2051	2052	2053	2054	2055	2056	2057	2058	2059	2060
$\frac{1}{2051}$	$\frac{1}{2052}$	$\frac{1}{2053}$	$\frac{1}{2054}$	$\frac{1}{2055}$	$\frac{1}{2056}$	$\frac{1}{2057}$	$\frac{1}{2058}$	$\frac{1}{2059}$	$\frac{1}{2060}$

$10000+\frac{10000-1}{10-1}=10000+\frac{9999}{9}=10000+1111=11111$; by [Rule 1.

$1+(10000-1)\times\frac{10}{9}=1+9999\times\frac{10}{9}=1+\frac{99990}{9}=1+11110$ $=11111$; by Rule 2.

The answer, in this case, embraces the money series, viz., 1 eagle, 1$, 1 dime, 1 cent, 1 mill. In the second operation the ratio is $\frac{1}{10}$, and its comp. $\frac{9}{10}$, which inverted makes $\frac{10}{9}$.

4. The extremes are 1 and 32, the ratio 2; what is the sum of the series? *Ans.* 63.

Here, if the ratio were $\frac{1}{2}$ it would not alter the result, since the extremes are given: the effect of it would be to change the direction of the series from the line of increase to that of decrease.

5. What is the sum of a series whose extremes are 1 and 81, or 81 and 1, and ratio 3? *Ans.* 121.

6. The extremes of a series are 1 and 256, and the ratio 4; what is the sum of all the terms? *Ans.* 341.

The class may draw out the preceding operations in the usual forms of addition, subtraction, multiplication, and division, without using the signs. g is not given in the annexed examples.

7. A series has 12 terms, the least is 1, the ratio 2; what are the greatest and sum of the series?

Ans. g2048, s 4095.

8. What would be the cost of a horse, if the price were estimated at 1 mill for the first shoe nail, 3 for the second, 9 for the third, 27 for the fourth, &c., in a series of 32 terms, and 3 the ratio? *Ans.* $926510094425.92.

In this example the ratio 3 is raised to the 10th power, and the 10th power cubed gives the 30th power; the 30th power$\times$3 gives the 31st, which is the number of terms less 1; then $3^{31}\times1=3^{31}$. This is the greatest term, and 1 is the least; add their difference, divided by ratio less 1.

9. A merchant sold 15 yards of satin, at 1 shilling for the first, 2 shillings for the second, 4 for the third, &c.; what did the satin amount to? *Ans.* 1638£ 7s.

10. A draper sold 20 yards of cloth, the first for 3d., the second for 9d., the third for 27d., &c., in a triple ratio; what is the whole amount? *Ans.* 21792402£ 10s. 1d.

11. The portion of a lady was $1.5 for the first month,

2061	2062	2063	2064	2065	2066	2067	2068	2069	2070
$\frac{1}{2061}$	$\frac{1}{2062}$	$\frac{1}{2063}$	$\frac{1}{2064}$	$\frac{1}{2065}$	$\frac{1}{2066}$	$\frac{1}{2067}$	$\frac{1}{2068}$	$\frac{1}{2069}$	$\frac{1}{2070}$

$3 for the second month, $6 for the third month, &c., in a double ratio, through the year; what was the sum?

Ans. $6142.5.

12. The least term is .004, the ratio 10, number of terms 8; what is the sum of the series?

Ans. 44444.444.

Note 1. When the least term is a unit, the sum of the series is found without the other extreme; thus, from the ratio involved to that power whose index is the number of terms, subtract the least term; then divide the remainder by the ratio less 1.

Formula. $(r^n - a) \div (r - 1) = s$.

Examples for the Slate.

1. Let a be 1, r 2, and n 5, to find s?

Ans. $(2^5 - 1) \div (2 - 1) = 31$.

2. Let a be 1, r 3, and n 5, to find s?

Ans. $(3^5 - 1) \div (3 - 1) = 121$.

3. Let a be 1, r 4, and n 5, to find s?

Ans. $(4^5 - 1) \div (4 - 1) = 341$.

4. The least term of a series is .0001, the ratio 10, and the number of terms 9; what is the sum of the series?

Ans. 11111.1111.

Note 2. To find the sum of a series decreasing *ad infinitum.*

RULE. Multiply the greatest term, viz. the given extreme, by the inverted complement of the ratio, and it is done.

Examples for the Slate.

1. Let g be 1, r $\frac{1}{2}$, (comp. $\frac{1}{2}$,) to find s?
2. Let g be 2, r $\frac{1}{3}$, (comp. $\frac{2}{3}$,) to find s?
3. Let g be 3, r $\frac{1}{4}$, (comp. $\frac{3}{4}$,) to find s?
4. Let g be 4, r $\frac{1}{5}$, (comp. $\frac{4}{5}$,) to find s?

Repeating and circulating decimals are series of this kind; in which $\frac{1}{10}$ is the ratio for a single repetend, $\frac{1}{100}$, the ratio for two repeating figures, $\frac{1}{1000}$ the ratio for a circulate of three figures, &c.; the complements of these are $\frac{9}{10}$, $\frac{99}{100}$, $\frac{999}{1000}$, &c., and these inverted make $\frac{10}{9}$, $\frac{100}{99}$, $\frac{1000}{999}$, &c.

2071	2072	2073	2074	2075	2076	2077	2078	2079	2080
$\frac{1}{2071}$	$\frac{1}{2072}$	$\frac{1}{2073}$	$\frac{1}{2074}$	$\frac{1}{2075}$	$\frac{1}{2076}$	$\frac{1}{2077}$	$\frac{1}{2078}$	$\frac{1}{2079}$	$\frac{1}{2080}$

5. What is the sum of a series whose greatest term is .3, or $\frac{3}{10}$, ratio $\frac{1}{10}$, and number of terms infinite? *Ans.* $\frac{1}{3}$.

6. What is the sum of a series whose greatest term is .4 or $\frac{4}{10}$, ratio $\frac{1}{10}$, and number of terms infinite? *Ans.* $\frac{4}{9}$.

7. What is the sum of a series whose greatest term is .6, or $\frac{6}{10}$, ratio $\frac{1}{10}$, and number of terms infinite? *Ans.* $\frac{2}{3}$.

8. The greatest term of a series is .24, or $\frac{24}{100}$, ratio $\frac{1}{100}$, and number of terms infinite; what is the sum? *Ans.* $\frac{8}{33}$.

9. The greatest term of a series is .428571, the ratio $\frac{1}{1000000}$, and number of terms infinite; what is the sum? *Ans.* $\frac{3}{7}$.

COMPOUND INTEREST.

Take p to represent principal, a amount, u rate per unum, and t time, as at page 131; or for t time take n number of payments. Also let r represent ratio; which for years will be $1+u$; for $\frac{1}{2}$ years, $1+\frac{1}{2}u$; for $\frac{1}{4}$ years, $1+\frac{1}{4}u$; &c.

And it is proper to remark that the terms of payment for which compound interest is to be computed, must be complete, without parts, or fractions of a term.

The columns of Table 1, following, are series in geometrical progression, the first term and ratio being the amount of $1 for 1 year, at the above rates. Higher powers may also be obtained by multiplying any two of these, or by squaring a power.

The years are the indices; and if a unit be divided by ratio, by r^2, by r^3, by r^4, &c., the quotients will be an extension of the same series below unity, showing the present worths of $1 hereafter due. This series is alluded to in division of decimals, and also in Involution: the method here proposed for forming a table of present worths, is however preferable to either of the others.

2081	2082	2083	2084	2085	2086	2087	2088	2089	2090
$\frac{1}{2081}$	$\frac{1}{2082}$	$\frac{1}{2083}$	$\frac{1}{2084}$	$\frac{1}{2085}$	$\frac{1}{2086}$	$\frac{1}{2087}$	$\frac{1}{2088}$	$\frac{1}{2089}$	$\frac{1}{2090}$

1. *Table showing* 20 *Powers of Various Ratios.*

Yrs.	At 3 per cent.	At 3½ per cent.	At 4 per cent.	At 4½ per cent.	At 5 per cent.	At 6 per cent.
1	1.03000	1.03500	1.04000	1.04500	1.05000	1.06000
2	1.06090	1.07123	1.08160	1.09202	1.10250	1.12360
3	1.09273	1.10872	1.12486	1.1412	1.15762	1.19102
4	1.12551	1.14752	1.16990	1.19252	1.21551	1.26250
5	1.15928	1.18768	1.21670	1.24620	1.27630	1.33823
6	1.19406	1.22925	1.26532	1.03230	1.34010	1.41852
7	1.22988	1.27228	1.31593	1.36086	1.40710	1.50363
8	1.26678	1.31681	1.36857	1.42210	1.47750	1.59385
9	1.30478	1.36289	1.42330	1.48610	1.55133	1.68948
10	1.34392	1.41059	1,48024	1.55300	1.62890	1.79085
11	1.38424	1.46000	1.53945	1.62290	1.71034	1.89830
12	1.42577	1.51107	1.60103	1.69590	1.79585	2.01220
13	1.46854	1.56360	1.66507	1.77200	1.88565	2.13293
14	1.51259	1.61870	1.73167	1.85200	1.97993	2.26900
15	1.55797	1.67535	1.80094	1.93530	2.07893	2.39656
16	1.60471	1.73400	1.87300	2.02240	2.18290	2.54035
17	1.65285	1.79470	1.94790	2.11340	2.29202	2.69277
18	1.70244	1.85750	2.02582	2.20850	2.40662	2.85434
19	1.75351	1.92250	2.10685	2.30786	2.52700	3.02560
20	1.80611	1.98980	2.19112	2.41171	2.65330	3.20714

Case 1. When the principal, rate and time are given to find the amount and compound interest.

RULE. Multiply that power of the ratio whose index is the number of years, ½ years, ¼ years, or other terms of payment, by the principal ; the product will be the amount; from which, to find the compound interest, subtract the principal.

Formula. $r^n \times p = a$; and $a - p =$ comp. interest.

Dem. Since r is the amount of $1 for 1 year, ½ year, or other term; and 1 is to its amount as p is to its amount; therefore

1. $1 : r^1 :: p : pr^1 =$ amt. of p for 1 term;
2. $1 : r^1 :: pr^1 : pr^2 =$ amt. of p for 2 terms;
3. $1 : r^1 :: pr^2 : pr^3 =$ amt. of p for 3 terms;

2091	2092	2093	2094	2095	2096	2097	2098	2099	2100
$\frac{1}{2091}$	$\frac{1}{2092}$	$\frac{1}{2093}$	$\frac{1}{2094}$	$\frac{1}{2095}$	$\frac{1}{2096}$	$\frac{1}{2097}$	$\frac{1}{2098}$	$\frac{1}{2099}$	$\frac{1}{2100}$

4. $1 : r^1 :: pr^3 : pr^4$=amt. of p for 4 terms; &c., to any number of terms required. Therefore $p\times$that power of r whose index is the number of terms equals the amount of p for the time, as prescribed in the rule

Examples for the Slate.

1. What is the amount of $200, for 3 years, at 5 per cent. per annum, compound interest?

r^3=1.05^3=1.15762, see table 1.
200 p

$231.52400 *Ans.*
Deduct 200.

$31.524 comp. interest.

2. What is the amount of 200 dollars for 5 years, at 5 per cent. per annum, compound interest? *Ans.* $255.26.

What do the following principals amount to, respectively, at the rates, and for the times set opposite, viz.:

3. $250 p, at 6 per c., for 4 years? *Ans.* $315.62½.
4. $300 p, at 5 per c., for 6 years? *Ans.* $402.03.
5. $400 p, at 4 per c., for 8 years? *Ans.* $547.428.
6. $450 p, at 4½ per c., for 10 years? *Ans.* $698.85.
7. $500 p, at 5 per c., for 12 years? *Ans.* $897.925.
8. $560 p, at 6 per c., for 14 years? *Ans.* $1270.64.
9. $640 p, at 6 per c., for 16 years? *Ans.* $1625.824.
10. $960 p, at 5 per c., for 18 yrs.? *Ans.* $2310.3552.
11. What is the compound interest of £720, for 3 years, at 5 per cent. per annum? *Ans.* £113 9s. 9d.
12. What is the compound interest of $720, for 7 years, at 4½ per cent. per annum? *Ans.* $259.82136.
13. What is the compound interest of a bond for £720, for 10 years, at 3½ per cent. per annum? *Ans.* £295 12s. 6d.
14. What will $1000 amount to in 20 years, at 6 per cent. per annum, compound interest? *Ans.* $3207.14.
15. What is the amount of $50 for 5 years, at 5 per

2101	2102	2103	2104	2105	2106	2107	2108	2109	2110
$\frac{1}{2101}$	$\frac{1}{2102}$	$\frac{1}{2103}$	$\frac{1}{2104}$	$\frac{1}{2105}$	$\frac{1}{2106}$	$\frac{1}{2107}$	$\frac{1}{2108}$	$\frac{1}{2109}$	$\frac{1}{2110}$

cent. per annum, compound interest, payable $\frac{1}{2}$ yearly ?
Ans. $64.00415.

Here are 10 terms of payment, and the ratio $1+\frac{1}{2}u=1+\frac{.05}{2}=1.025$; which, involved to the 10th power, will be the amount of $1 for the given time. Contract the operation to 6 decimal places. See ex. 5, page 142.

16. What is the compound interest of $500, payable $\frac{1}{4}$ yearly, for 5 years, at 5 per cent. per annum ?
Ans. $141.018.

Here are 20 terms of payment, and the ratio $1+\frac{1}{4}u=1+\frac{.05}{4}=1.0125$; and $1.0125^{20}\times500=a$; then $a-p=$comp. interest.

Case 2. When the amount, rate, and time are given, to find the principal or present-worth.

Rule. Divide the whole amount by the ratio involved to the time ; that is, involved to the power whose index is the number of years, $\frac{1}{2}$ years, $\frac{1}{4}$ years, or other terms ; the quotient will be the present worth.

Formula. $a\div r^n=p$, or present worth.

Note. For powers of the annual ratio see Table 1.

Examples for the Slate.

1. What sum, in hand, is equivalent to $231.524 at the end of 3 years, discounting at 5 per cent. per annum ?
Ans. $200.

Ratio $^3=1.05^3=1.15762$, the amount of $1.
1.15762)231.52400(200 *p*, or *p. w.*

What are the present worths of the amounts mentioned below, discounting at the rates, and for the times set opposite, respectively, at compound interest, viz. :

2. $255.26, for 5 yrs., at 5 per cent. ? *Ans.* $200.
3. 315.62\frac{1}{2}$, for 4 yrs., at 6 per cent. ? *Ans.* $250.
4. $402.03, for 6 yrs., at 5 per cent. ? *Ans.* $300.
5. $547.428, for 8 yrs., at 4 per cent. ? *Ans.* $400.
6. $698.85, for 10 yrs., at 4$\frac{1}{2}$ per cent. ? *Ans.* $450.
7. $897.925, for 12 yrs., at 5 per cent. ? *Ans.* $500.
8. $1270.64, for 14 yrs., at 6 per cent. ? *Ans.* $560.
9. $1625.824, for 16 yrs., at 6 per cent. ? *Ans.* $640.

2111	2112	2113	2114	2115	2116	2117	2118	2119	2120
$\frac{1}{2111}$	$\frac{1}{2112}$	$\frac{1}{2113}$	$\frac{1}{2114}$	$\frac{1}{2115}$	$\frac{1}{2116}$	$\frac{1}{2117}$	$\frac{1}{2118}$	$\frac{1}{2119}$	$\frac{1}{2120}$

10. $2310.3552, for 18 yrs., at 6 per cent.? *Ans.* $960.

11. What is the discount on $833.4875, for 3 years, at 5 per cent. per annum, compound interest?
Ans. $113.4875.

12. How much is the discount on $979.82136, for 7 years, at 4½ per cent. per annum, compound interest?
Ans. $259.82136.

Case 3. When the amount, principal, and time are given, to find the ratio.

RULE. Divide the amount by the principal, and extract such root of the quotient as the number of years indicate.

Note. In the table of powers of ratios, opposite the years, if not more than 20, will be found the quotient; and at the head of the column will be found the root, which is the ratio sought. The laborious operation of extracting roots of high powers, is thus superseded by preparing a table of the powers of ratios.

Formula. $\sqrt[n]{a \div p} = r$; because $r^n = a \div p$.

Examples for the Slate.

1. If 315£ 12s. 6d. be the amount of 250£ at compound interest for 4 years, what is the ratio? or, what is the amount of 1£ for 1 year?

```
 12)  6.0
 20) 12.5
250)315.625(1.2625
```

Then $\sqrt[4]{1.2625} = 1.06$ *Ans.*

When the amount a, principal p, and time n, are given as in the examples below, what are the amounts of $1 for 1 year respectively? or what are the ratios?

2. $231.524 a, $200 p, 3 yrs.? *Ans.* 1.05.
3. $255.26 a, $200 p, 5 yrs.? *Ans.* 1.05.
4. $315.62½ a, $250 p, 4 yrs.? *Ans.* 1.06.
5. $402.03 a, $300 p, 6 yrs.? *Ans.* 1.05.
6. $547.428 a, $400 p, 8 yrs.? *Ans.* 1.04.
7. $698.85 a, $450 p, 10 yrs.? *Ans.* 1.045.
8. $897.925 a, $500 p, 12 yrs.? *Ans.* 1.05.

2121	2122	2123	2124	2125	2126	2127	2128	2129	2130
$\frac{1}{2121}$	$\frac{1}{2122}$	$\frac{1}{2123}$	$\frac{1}{2124}$	$\frac{1}{2125}$	$\frac{1}{2126}$	$\frac{1}{2127}$	$\frac{1}{2128}$	$\frac{1}{2129}$	$\frac{1}{2130}$

9. $1270.64 *a*, $560 *p*, 14 yrs.? *Ans.* 1.06.
10. $1625.824 *a*, $640 *p*, 16 yrs.? *Ans.* 1.06.
11. $2310.3552 *a*, $960 *p*, 18 yrs.? *Ans.* 1.06.
12. $833.4875 *a*, $720 *p*, 3 yrs.? *Ans.* 1.05.

13. A bond for £125 7s. 6d., being at interest for 7 years, amounted to £176 8s. 3¾d.; how much is the ratio or amount of £1, for 1 year? *Ans.* 1.05.

14. If the compound interest of $100, for 10 years, be $55.3, what will have been the ratio? *Ans.* 1.045.

Case 4. When the amount, principal and rate are given to find the time.

Rule. Divide the amount by the principal, and involve the ratio to that power which shall be equal to the quotient; the index of the power will be the time.

Or, find the quotient under the rate, in Table 1; and the years stand opposite on the left.

Formula. $a \div p = r^n$.

Examples for the Slate.

1. In what time will 250£ amount to 315£ 12s. 6d. computing compound interest at 6 per cent. per annum?

6.0 d. *Ans.* 4 years.
12.5 s.
£315.625÷250=1.2625. See Table 1.
Then $r^n = 1.06^4 = 1.2625$. Therefore $n=4$ years.

When the amount *a*, principal *p*, and rate (and therefore ratio), are given, as in the annexed examples, what are the times, respectively?

2. $231.524 *a*, $200 *p*, 1.05 *r*? *Ans.* *n* 3 yrs.
3. $255.26 *a*, $200 *p*, 1.05 *r*? *Ans.* *n* 5 yrs.
4. $315.625 *a*, $250 *p*, 1.06 *r*? *Ans.* *n* 4 yrs.
5. $402.03 *a*, $300 *p*, 1.05 *r*? *Ans.* *n* 6 yrs.
6. $547.428 *a*, $400 *p*, 1.04 *r*? *Ans.* *n* 8 yrs.
7. $698.85 *a*, $450 *p*, 1.045 *r*? *Ans.* *n* 10 yrs.
8. $897.925 *a*, $500 *p*, 1.05 *r*? *Ans.* *n* 12 yrs.
9. $1270.64 *a*, $560 *p*, 1.06 *r*? *Ans.* *n* 14 yrs.
10. $1625.824 *a*, $640 *p*, 1.06 *r*? *Ans.* *n* 16 yrs.

2131	2132	2133	2134	2135	2136	2137	2138	2139	2140
$\frac{1}{2131}$	$\frac{1}{2132}$	$\frac{1}{2133}$	$\frac{1}{2134}$	$\frac{1}{2135}$	$\frac{1}{2136}$	$\frac{1}{2137}$	$\frac{1}{2138}$	$\frac{1}{2139}$	$\frac{1}{2140}$

11. $2310.3552 *a*, $960 *p*, 1.06 *r* ? *Ans. n* 18 yrs.
12. $833.4875 *a*, $720 *p*, 1.05 *r*? *Ans. n* 3 yrs.
13. In what time will £125 7s. 6d. amount to £176 8s. $3\frac{3}{4}$d.; at 5 per cent. per annum? *Ans.* 7 yrs.
14. In what time will the compound interest of $100 arise to $55.3, at $4\frac{1}{2}$ per cent. per annum?

Ans. 10 years.

ANNUITIES AT COMPOUND INTEREST.

Annuities are periodical incomes, arising from money lent, or from houses, lands, salaries, pensions, &c., payable for the most part, half or quarter yearly.

Annuities in possession are such as have commenced.

Annuities in reversion are such as will not begin until some event shall have happened, or until some time shall have elapsed.

Perpetuities are perpetual annuities.

Annuities are said to be in arrear when the payments are not made for several terms.

The *amount* of an annuity is the sum arising from the addition of several payments, including the compound interest of each, for the time which it may be in arrear.

The present worth of an annuity is the sum which ought to be given for it, supposing it to be bought off at once.

Case 1. When the annuity, time and rate are given to find the amount.

Rule. Find the sum of a series in geometrical progression, whose least term is 1, number of terms 1 less than the number of payments due, and whose ratio is the amount of $1, or £1 for 1 term of payment; multiply this sum by the given annuity for the answer.

2141	2142	2143	2144	2145	2146	2147	2148	2149	2150
$\frac{1}{2141}$	$\frac{1}{2142}$	$\frac{1}{2143}$	$\frac{1}{2144}$	$\frac{1}{2145}$	$\frac{1}{2146}$	$\frac{1}{2147}$	$\frac{1}{2148}$	$\frac{1}{2149}$	$\frac{1}{2150}$

Note. The sum of the series mentioned in this rule is the amount of annuity of $1, or £1 for the time; which may also be found thus: $1 being the sum due at the end of the first term, to 1 add the ratio; this will be the amount of the annuity at the end of the 2d term: to this sum add ratio^2; this will be the amount of the annuity at the end of the 3d term: to this sum add ratio^3; this will be the amount of the annuity at the end of the 4th term, &c. Thus, from Table 1, compound interest, is the following table constructed.

2. *Table showing the Amounts of an Annuity of* 1£ *or* $1.

Yrs.	At 3 per cent.	At 3½ per cent.	At 4 per cent.	At 4½ per cent.	At 5 per cent.	At 6 per cent.
1	1.	1.	1.	1.	1.	1.
2	2.03	2.035	2.04	2.045	2.05	2.06
3	3.9009	3.10623	3.1216	4.13702	3.1525	3.1836
4	4.18363	4.21495	4.24646	4.27822	4.31012	4.37462
5	5.30914	5.36247	5.41636	5.47074	5.52563	5.63712
6	6.46842	6.55015	6.63306	6.71694	6.80193	6.97535
7	7.66248	7.7794	7.89838	8.01924	8.14203	8.39387
8	8.89236	9.05168	9.21431	9.3801	9.54913	9.8975
9	10.15914	10.36849	10.58288	10.80222	11.02663	11.49135
10	11.46392	11.73138	12.00618	12.28822	12.57796	13.18083
11	12.80784	13.14197	13.48642	13.84122	14.20686	14.97168
12	14.19208	14.60197	15.02587	15.46412	15.9172	16.86998
13	15.61785	16.11304	16.6269	17.16	17.71305	18.88218
14	17.08639	17.67664	18.29197	18.9322	19.5987	21.01511
15	18.59898	19.29534	20.02364	20.7842	21.57863	23.27601
16	20.15695	20.97069	21.82458	22.7195	23.65756	25.67267
17	21.76166	22.70469	23.69758	24.7419	25.84046	28.21302
18	23.41451	24.49939	25.64548	26.8553	28.13248	30.90579
19	25.11695	26.35689	27.6713	29.0638	30.5391	33.76013
20	26.87046	28.27939	29.77815	31.37166	33.0661	36.78573
21	28.67657	30.26919	31.96927	33.78366	35.7194	39.99287

Examples for the Slate.

1. An annuity of $100 per annum being 7 years in ar-

2151	2152	2153	2154	2155	2156	2157	2158	2159	2160
$\frac{1}{2151}$	$\frac{1}{2152}$	$\frac{1}{2153}$	$\frac{1}{2154}$	$\frac{1}{2155}$	$\frac{1}{2156}$	$\frac{1}{2157}$	$\frac{1}{2158}$	$\frac{1}{2159}$	$\frac{1}{2160}$

rear, what is the amount due, at 6 per cent., compound interest?

$$r^{n-1}\times a=g=\overline{1.06}^{7-1}\times 1=\overline{1.06}^{6}\times 1.41852.$$

Then 1.41852 greatest term.
1. least term.

$r-1=1.06-1=.06$) .41852 dif. extremes.
divided by $r-1=$ 6.97533+
1.41852 greatest term.

sum of \$1 annuity= 8.39385 see Table 2.
100

\$839.38500 *Ans.*

2. If an annual rent of \$84 remain unpaid for 11 years, bearing compound interest at 4½ per cent. per annum, what will be the amount? *Ans.* \$1162.662+.

3. If an annuity of £30 per annum remain unpaid 15 years, what then is the amount, at 6 per cent. per annum, compound interest? *Ans.* £647 7s. 2d.

4. What will an annual rent of \$200 per annum amount to in 10 years, allowing compound interest at 6 per cent. per annum? *Ans.* \$2636.166.

5. What is the amount of an annuity of \$28, for 21 years, with compound interest at 5 per cent.? *Ans.* \$1000.143.

Case 2. To find the present worths of annuities in possession, when the yearly, ½ yearly, or ¼ yearly sum, rate of interest, and time are given.

RULE. Find the present worths of the first and last terms by Case 2, compound interest, for the greatest and least terms of a series in geometrical progression; the sum of this series will be the present worth of the annuity: the ratio will be the amount of \$1 at the given rate and time.

2161	2162	2163	2164	2165	2166	2167	2168	2169	2170
$\frac{1}{2161}$	$\frac{1}{2162}$	$\frac{1}{2163}$	$\frac{1}{2164}$	$\frac{1}{2165}$	$\frac{1}{2166}$	$\frac{1}{2167}$	$\frac{1}{2168}$	$\frac{1}{2169}$	$\frac{1}{2170}$

Note. The present worth of $1, as of any sum, is the quotient of 1 divided by the ratio involved to the power whose index is the number of terms, that is, for 1 term divide by r^1, for 2 terms divide by r^2, for 3 terms divide by r^3, &c. Or, divide 1 by r, and the quotient by r, and this quotient by r, and so on continually; the quotients will be the present worths for the respective terms; which are evidently a decreasing geometrical series, and from which is formed the annexed table: thus, the terms being found by division as above, and as prescribed in division of decimals, the 1st number in each column is the quotient of 1 by r; the 2d number is the 1st+the 2d quotient: the 3d number is the 2d+the 3d quotient; the 4th number is the 3d+the 4th quotient, &c.

Having thus the present worth of a $1 annuity, it may be multiplied by the given annuity for the answer.

3. *Table showing the Present Worth of £1 or $1 Annuity.*

Yrs.	At 3 per cent.	At 3½ per cent.	At 4 per cent.	At 4½ per cent.	At 5 per cent.	At 6 per cent.
1	.97087	.96618	.96154	.95694	.95238	.9434
2	1.91347	1.89969	1.88609	1.8727	1.85941	1.8334
3	2.82862	2.80163	2.77509	2.74876	2.72325	2.67301
4	3.71712	3.67307	3.6299	3.58752	3.54595	3.4651
5	4.57974	4.51504	4.45181	4.39	4.32948	4.21236
6	5.41724	5.32853	5.24212	5.1579	5.07569	4.91732
7	6.23034	6.11451	6.00205	5.8927	5.78637	5.58238
8	7.01976	6.87391	6.73271	6.5959	6.46321	6.2098
9	7.78618	7.60763	7.43533	7.2688	7.10782	6.8017
10	8.53028	8.31653	8.11089	7.91272	7.72173	7.3601
11	9.2527	9.00146	8.76049	8.52892	8.30641	7.8869
12	9.95409	9.66322	9.3851	9.11858	8.86325	8.38384
13	10.63504	10.3026	9.9857	9.68285	9.39357	8.85268
14	11.29616	10.9204	10.56312	10.22282	9.89864	9.29498
15	11.93803	11.51727	11.1184	10.73955	10.37965	9.71225
16	12.5612	12.09395	11.6523	11.23401	10.83777	10.1059
17	13.1662	12.65113	12.1657	11.70719	11.27407	10.47726
18	13.7536	13.48947	12.6593	12.16	11.68959	10.8276
19	14.32389	13.7096	13.13394	12.5933	12.08532	11.16812
20	14.87758	14.41214	13.59032	13.00793	12.46221	11.46992
21	15.41514	14.69768	14.02916	13.40472	13.82115	11.76408

2171	2172	2173	2174	2175	2176	2177	2178	2179	2180
$\frac{1}{2171}$	$\frac{1}{2172}$	$\frac{1}{2173}$	$\frac{1}{2174}$	$\frac{1}{2175}$	$\frac{1}{2176}$	$\frac{1}{2177}$	$\frac{1}{2178}$	$\frac{1}{2179}$	$\frac{1}{2180}$

Examples for the Slate.

1. How much ready money should be paid for an annuity of $300, to continue 7 years, allowing compound interest upon the advance, at 6 per cent. per annum?

Here $a \div r^1 = p.\ w.$ for the first year. greatest.
and $a \div r^7 = p.\ w.$ for the seventh year. least.
$r=1.06$; therefore 1.06)300.00(283.019 greatest.
and $\overline{1.06}^7=1.50363$)300.00000(199.517 least.

$r-1=1.06-1=$ .06) 83.502

Difference of extremes$\div(r-1)=$ 1391.7
greatest + 283.019

$1674.719 *Ans.*

The amount of $1=5.58238. See Table 3.
Therefore, 5.58238×300=$1674.714; nearly the same.

2. What is the value in cash of a lease worth 50£ per annum, there being 10 years unexpired, allowing compound interest at 4 per cent.? *Ans.* 405£ 10s. 10½d.

3. An annuity of $200 per annum, to continue 15 years, is to be disposed of for ready money, on which a discount of 4½ per cent. per annum will be allowed; what sum would purchase this? *Ans.* $2147.91.

4. What is the present worth of an annuity of $800 for 21 years, at a discount of 5 per cent. per annum? *Ans.* $11056.92.

Case 3. To find the present worth of annuities in reversion.

Rule. Find the present worth of the annuity as if it were now commencing; then find the present worth of that sum, payable at the time of reversion, by Case 2, Compound Interest.

Examples for the Slate.

1. What is the value now, of an annuity of 20£, which

2181	2182	2183	2184	2185	2186	2187	2188	2189	2190
$\frac{1}{2181}$	$\frac{1}{2182}$	$\frac{1}{2183}$	$\frac{1}{2184}$	$\frac{1}{2185}$	$\frac{1}{2186}$	$\frac{1}{2187}$	$\frac{1}{2188}$	$\frac{1}{2189}$	$\frac{1}{2190}$

shall commence 2 years hence, and continue 4 years, discounting at 5 per cent. per annum?

The *p. w.* of 1£=3.54595. See Table 3, opposite 4 years, and under 5 per cent.

Then 3.54595 pres. worth of 1£.
×20

$\overline{1.05}^2$=1.1025)70.91900(64.326 nearly.

2. What should be paid in cash for an annuity of 24£, to begin 7 years hence, and to continue 21 years, allowing compound interest upon the advance, at 6 per cent. per annum? *Ans.* £187.77+.

3. What should be paid for an annuity of $800, which shall commence in 2 and continue 7 years; discounting at 5 per cent. per annum? *Ans.* $4198.726.

PERPETUITIES.

When the terms of payment are infinite, the remote annuities are inaccessible upon the principles of involution. To remedy this inconvenience erroneous principles are admitted into these calculations; among which is the supposition that the annuity is annihilated upon being divided by the ratio infinitely involved; that is, that $a \div r^n = 0$, when n represents infinitude. But all this is foreign to the subject, and our conceptions of the present worth of perpetuities must be derived from other *data.*

It will be equitable, therefore, that the annuity yield to the purchaser an equivalent for the interest of his money, regularly improved; and we may conceive the annuity itself to be this equivalent; whence we have in

Case 1. The rate per centum or per unum, and the periodical sum to find the present worth of perpetuities in possession.

2191	2192	2193	2194	2195	2196	2197	2198	2199	2200
$\frac{1}{2191}$	$\frac{1}{2192}$	$\frac{1}{2193}$	$\frac{1}{2194}$	$\frac{1}{2195}$	$\frac{1}{2196}$	$\frac{1}{2197}$	$\frac{1}{2198}$	$\frac{1}{2199}$	$\frac{1}{2200}$

RULE. Divide the annuity by the rate per unum, and the work is done.

For, since the annuity is taken as the interest of the sum advanced; if we assume the rate per unum as .05, and the annuity as $50, it will be .05 : 1 : : 50.00 : 1000; wherefore the quotient of the annuity divided by the rate per unum is the present worth.

Examples for the Slate.

1. What sum in ready money should be paid for an estate worth $1200 per annum, allowing the purchaser 6 per cent. per annum upon his investment?
Ans. 1200.00÷.06=20000 dollars.

2. What is the present worth of a freehold estate of $500 per annum, allowing the purchaser 5 per cent. per annum? *Ans.* $10000.

3. What should be paid in cash for an annuity of £184, allowing 4½ per cent. per annum upon the purchase money? *Ans.* £4088 17s. 10⅔d.

Note. When the payments are ½, or ¼ yearly, the value of each annuity will be enhanced, in the ratio of the simple yearly interest to the compounded ½ or ¼ yearly interest.

RULE. As the yearly rate is to the annuity so is the compounded ½ or ¼ yearly rate to the enhanced annuity. Divide this sum by the rate per unum, as before.

Table 4. Compounded ½ and ¼ yearly rates per cent.

yearly.	½ yearly.	¼ yearly.
3.	3.0225	3.0339
3.5	3.5307	3.5461
4.	4.04	4.0604
4.5	4.5506	4.5765
5.	5.0625	5.0945
5.5	5.5757	5.6143
6.	6.09	6.1364

In constructing the ½ yearly rates, add to the yearly rate the interest of its half for 6 months. For the ¼ yearly rates add to the yearly rate the interest of its ¼ part for 9 months, and also the interest of its ¼ part for 6 months, *and the same for* 3 months, at the given rate per cent.

2201	2202	2203	2204	2205	2206	2207	2208	2209	2210
$\frac{1}{2201}$	$\frac{1}{2202}$	$\frac{1}{2203}$	$\frac{1}{2204}$	$\frac{1}{2205}$	$\frac{1}{2206}$	$\frac{1}{2207}$	$\frac{1}{2208}$	$\frac{1}{2209}$	$\frac{1}{2210}$

Examples for the Slate.

1. What is the present worth of $200 per annum, forever, payable ½ yearly, allowing 4 per cent. to the purchaser?

4 : 4.04 : : 200 : 202, enhanced annuity.
Then, .04 : 1 : : 202 : 5050, present worth.

2. Required the present value of $250 per annum, forever, payable half yearly, allowing the purchaser 5 per cent. per annum? *Ans.* $5062.5.

3. What sum in cash should be paid for a perpetuity of $300, payable ¼ yearly, discount being allowed at 6 per cent.? *Ans.* $5113.66⅔.

Case 2. To find the present worth of perpetuities in reversion.

Rule. Find the present worth of the perpetuity as if it were now commencing, by Case 1; then find the present worth of that sum, payable at the time of reversion, by Case 2, compound interest.

Examples for the Slate.

1. What is the present worth of a perpetual annuity, to commence at the end of 3 years, the annual sum being £27, and the discount 6 per cent.?

$27 \div .06 = 450$£, commencing 3 yrs. hence.
Then $450 \div 1.06^3 = 450 \div 1.19102 = £377.8274$ *Ans.*

2. What is the present worth of an annuity of £60, commencing at the end of 2 years, and continuing forever, allowing discount at 6 per cent.? *Ans.* £889 19s. 11¾d.

3. What is the present worth of a perpetuity of $500, commencing 4 years hence, allowing the purchaser 4 per cent. per annum? *Ans.* $8227.

4. What is the present worth of a perpetuity of $1000, payable ½ yearly, commencing 4 years hence, allowing the purchaser 6 per cent. per annum? *Ans.* $13399.21+.

2211	2212	2213	2214	2215	2216	2217	2218	2219	2220
$\frac{1}{2211}$	$\frac{1}{2212}$	$\frac{1}{2213}$	$\frac{1}{2214}$	$\frac{1}{2215}$	$\frac{1}{2216}$	$\frac{1}{2217}$	$\frac{1}{2218}$	$\frac{1}{2219}$	$\frac{1}{2220}$

ALLIGATION, OR LINKING

This rule takes its name from its mode of operation: it combines simple quantities proportionally in a composition, and resolves a mean rate into the rates of its simple ingredients.

Alligation is usually divided into four cases: viz., *medial*, *alternate*, *partial*, and *total*, according to the character of the parts given and required.

Ingredients of comparatively small cost, as water with wine, or alloy with gold or silver, are considered free, and represented in the work by a 0.

The fineness of silver and gold is estimated by carats and 32d parts of a carat. A carat is $\frac{1}{24}$ part of the mass.

The coins of the United States are alloyed $\frac{1}{10}$; viz., the silver with copper, and the gold with silver and copper in equal parts. Their fineness is therefore very nearly 21 carats and $\frac{19}{32}$ of a carat.

An ounce troy of pure gold is worth	\$20.672
An ounce troy of pure silver is worth	1.292
An ounce troy of U. S. coined gold is worth	18.60$\frac{20}{43}$
An ounce troy of U. S. coined silver is worth	1.16$\frac{12}{43}$

Medial is the term used when the mean rate is sought and the quantities and rates of the simples given.

RULE. As the sum of the simple quantities is to their entire value at the given rates, so is any part of the composition to its value.

Examples for the Slate.

1. The silver coins mentioned below are set down at their price by the ounce, in Federal money; and it is required to find how much an ounce of the mass would be worth, if all were combined in equal quantities?

Ans. \$1.06$\frac{28}{41}$.

Old U. S. coins,	\$1.15	Sardinian coins,	1.18
New U. S. coins,	1.16	Crown of Brabant,	1.13

2221	2222	2223	2224	2225	2226	2227	2228	2229	2230
$\frac{1}{2221}$	$\frac{1}{2222}$	$\frac{1}{2223}$	$\frac{1}{2224}$	$\frac{1}{2225}$	$\frac{1}{2226}$	$\frac{1}{2227}$	$\frac{1}{2228}$	$\frac{1}{2229}$	$\frac{1}{2230}$

Pieces of 640 reis of Brazil,	1.17
Piece of 800 reis of Brazil,	1.18
Columbian Dollar,	.94
Dollar of Grenada,	.87
¼ Dols. Bolivia,	.85
½ and ¼ Dols. Peru,	.84
Pistareens, Spanish,	1.05
Pieces of 200 reis Portuguese,	1.19
British coins,	1.20
French do.,	1.18
Florin, Brunswick,	1.29
Specie Thaler,	1.07½
Thaler,	.97
Hesse specie Thaler,	1.07½
Belgic Crown,	1.13
do. 5 francs,	1.15½
do. Rix Dollar,	1.13
do. do.	1.13½
Austrian Rix Dollars and halves, or Florins,	1.08
20 Kreutzers,	.75
10 Kreutzers,	.64½
Scudo and ½ Lira,	1.16
Thalers, Prussia,	.97
Halves,	.86
Specie Thaler, Saxony,	1.08
Thaler, Saxony,	.97
Specie Thaler, Baden,	1.07½
Crown, do.	1.13
10 Kreutzers,	.64½
Gulden,	1.16
Netherlands Rix Dollar,	1.18
Three Florins piece,	1.16
Norway, Sweden and Denmark specie Dol. 60 Schillings,	1.13
Rigsbank Dollar,	1.13½
Rouble, Russia,	1.13
10 Pauls, Tuscany,	1.18½
10 and 5 Livre pieces, Tuscany,	1.24½

The columns above contain the prices of 41 ounces of coined silver; and if the entire sum be divided by the number of ounces, the quotient will be the mean rate required, or the price of a single ounce of the composition. If equimultiples, or equal parts of the quantities were taken the result would be the same; but it is very easily seen that other combinations might be made which would vary the result. The coins not mentioned above are for the most part of the value of 116 cents an ounce.

2. If 16 bushels of wheat at \$1.5, 24 bushels of rye at \$1.25, and 32 bushels of maize at \$1 per bushel, be mixed together, what will a bushel of the mixture be worth?

bu.		\$				
16	at	1.5 = 24				
24	at	1.25 = 30				
32	at	1. = 32				
—		—		bu.		
72	:	86	: :	1	:	1.194+ *Ans.*

2231	2232	2233	2234	2235	2236	2237	2238	2239	2240
$\frac{1}{2231}$	$\frac{1}{2232}$	$\frac{1}{2233}$	$\frac{1}{2234}$	$\frac{1}{2235}$	$\frac{1}{2236}$	$\frac{1}{2237}$	$\frac{1}{2238}$	$\frac{1}{2239}$	$\frac{1}{2240}$

3. A grocer mixes 3 parcels of tea together, viz.: 4 lbs. at \$$1\frac{1}{4}$, 7 lbs. at \$1.4, and 9 lbs. at \$$1\frac{1}{2}$ per lb.; how much is the mixture worth per lb.? *Ans.* \$1.415.

4. A tobacconist mixes 36 lbs. of leaf at 18 cents, 12 lbs. at 24 cents, and 12 lbs. at 22 cents; and desires to know how much the composition is worth per lb.? *Ans.* 20 cents.

5. A goldsmith mixes 4 lbs. of gold 20 carats fine, with 5 lbs. of 18 carats fine, and 6 lbs. of 16 carats fine; how many carats fine will this composition be? *Ans.* $17\frac{11}{15}$ carats.

Alternate quantities of the simples are taken when the rates of the simples and of the composition are given.

Rule. Set the rates of the simples in such order that each rate which is less may be connected with one or more that is greater than the mean rate, and the contrary: write the difference between the mean rate and the rate of each simple opposite the rate with which the latter is connected: then if but one difference stand against any rate, it is the quantity; but if more, their sum is the quantity to be taken of that rate.

Examples for the Slate.

1. How much wine at \$1.5, \$2.25, and \$2.5 per gallon, should be taken to make a mixture worth \$2 per gallon?

Mean rate 2\$. { 1.5, 2.25, 2.5 } .25+.5=.75 at \$1.5; .5 at \$2.25; .5 at \$2.5 } *Ans.*

2. How much sugar at 8 cents, and at $6\frac{1}{4}$ cents per lb. should be taken into a composition worth $7\frac{1}{2}$ cents per lb.? *Ans.* $1\frac{1}{4}$ lb. at 8 cts., and $\frac{1}{2}$ lb. at $6\frac{1}{4}$ cts.

3. How much of each of the following coins should be taken into a composition of 20 carats fine, viz.:

of Peruvian $\frac{1}{2}$ dollars, $15\frac{23}{32}$ carats? *Ans.* 87 parts.
of Bolivian dollars, 16 " ? *Ans.* 87 parts.
of Spanish pistareens, $19\frac{16}{32}$ " ? *Ans.* 87 parts.

2241	2242	2243	2244	2245	2246	2247	2248	2249	2250
$\frac{1}{2241}$	$\frac{1}{2242}$	$\frac{1}{2243}$	$\frac{1}{2244}$	$\frac{1}{2245}$	$\frac{1}{2246}$	$\frac{1}{2247}$	$\frac{1}{2248}$	$\frac{1}{2249}$	$\frac{1}{2250}$

Brazil piece of 800 reis, $21\frac{11}{32}$ car. ? *Ans.* 281 parts.
" 1200 reis, $21\frac{12}{32}$ " ? *Ans.* 281 parts.

4. How much of each of the following coins should be taken into a composition of $21\frac{16}{32}$ carats fine, viz.:
U. S. old coins of $21\frac{13}{32}$ carats? *Ans.* 5 parts.
Mexican coins of $21\frac{18}{32}$ carats? *Ans.* 364 parts.
U. S. new coins of $21\frac{12}{32}$ carats? *Ans.* 364 parts.
Peruvian $\frac{1}{2}$ dollars of $15\frac{23}{32}$ carats? *Ans.* 5 parts.
Bolivian $\frac{1}{2}$ dollar of 16 carats? *Ans.* 5 parts.

5. How much of each of the following coins should be taken into a composition of 21 carats fine? viz.:
British coins of $22\frac{10}{32}$ carats fine? *Ans.* 320 parts.
French coins of $21\frac{28}{32}$ carats fine? *Ans.* 320 parts.
Austrian rix dollar of 20 carats fine? *Ans.* 70 parts.
Crown of Brabant of 21 carats fine? *Ans.* More or less.
Ten kreutzer of 12 carats fine? *Ans.* 70 parts.

6. How many Prussian thalers of 18 carats, *Ans.* 227.
" " $\frac{1}{2}$ thalers of 16 carats, *Ans.* 227.
" Brunswick florins of 24 car., *Ans.* 192.
" Italian 5 livres of $23\frac{3}{32}$ car., *Ans.* 192.
will unite in a mass of 20 carats fine?

Note. When the simple quantities do not exceed *three*, there can be but one way of linking them; but various methods of linking may be used when the simples are numerous. Multiples or parts of the numbers found by linking will answer the question.

Partial alligation furnishes the quantity of one of the simples, the rates of all the simples, and also the mean rate, and requires the quantities of the other simples.

RULE. Find the differences by linking, as in alligation alternate; then the difference opposite to the given rate is to its quantity as each of the other differences is to its quantity.

Examples for the Slate.

1. A jeweler has 10 ounces of gold of 16 carats fine, which he would mix with gold of 18, 22, and 24 carats

2251	2252	2253	2254	2255	2256	2257	2258	2259	2260
$\frac{1}{2251}$	$\frac{1}{2252}$	$\frac{1}{2253}$	$\frac{1}{2254}$	$\frac{1}{2255}$	$\frac{1}{2256}$	$\frac{1}{2257}$	$\frac{1}{2258}$	$\frac{1}{2259}$	$\frac{1}{2260}$

fine ; how much of each of the latter kinds must be taken to make the mass 20 carats fine ?

Mean rate 20 { 16, 18, 22, 24 }

16 — 2+4=6 ; therefore, since the differences are equal, the quantities will be equal ; for
18 — 2+4=6
22 — 2+4=6
24 — 2+4=6

6 : 10 : : 6 : 10. *Ans.*

But a more simple combination may be made, viz. :

Mean rate 20 { 16, 18, 22, 24 }

16 — 4
18 — 2
22 — 2
24 — 4

And, since the given quantity at 16 carats fine is 10 oz., it will be 10 oz. of 24 carats, and 5 oz. of 18 and 22.

2. How many ounces of Austrian kreutzers of 12 carats fine, of Prussian $\frac{1}{2}$ thalers of 16 carats, of Prussian thalers of 18 carats, and of Austrian rix dollars of 20 carats, should be mixed with 4 ounces of Brunswick florins of 24 carats, to make a mass of 19 carats fine ?

Ans. 2 oz. of kreutzers, 2 oz. of $\frac{1}{2}$ thalers, 8 dwts. of thalers, and 8 dwts. of rix dollars, or any quantities in the same ratio.

3. How many dwts. and grs. of the doubloons—
of Mexico $20\frac{24}{32}$ carats, of Popaya $20\frac{19}{32}$ carats,
of La Plata $19\frac{18}{32}$ carats, of Lima $20\frac{26}{32}$ carats,
of Chili $20\frac{30}{32}$ carats, of Cuzco $20\frac{29}{32}$ carats,
of Bogota $20\frac{28}{32}$ carats, should be taken with 16 dwts. of pure gold 24 carats fine, to make a mass of 22 carats fine ? *Ans.* 3 dwts. $8\frac{16}{51}$ grs. of each.

4. The fineness in carats of certain gold coins is as follows, viz. :

British Guineas, &c., $21\frac{31}{32}$, French Louis d'or, $21\frac{21}{32}$,
Portuguese Moidore, $21\frac{29}{32}$, do. 40 francs, $21\frac{18}{32}$,
do. Johannes, $21\frac{28}{32}$, Austrian Ducat, $23\frac{20}{32}$,
Spanish Pistole, $20\frac{30}{32}$, do. Sovereign, 22.

How many dwts. of these must be mixed with 60 dwts. of pure gold to make a mass of $22\frac{24}{32}$ carats fine ?

Ans. 60 dwts. of the Austrian ducat, and 17 dwts. $8\frac{32}{47}$ grs. of each of the other pieces.

5. How many dwts. of each of the following coins,

2261	2262	2263	2264	2265	2266	2267	2268	2269	2270
$\frac{1}{2261}$	$\frac{1}{2262}$	$\frac{1}{2263}$	$\frac{1}{2264}$	$\frac{1}{2265}$	$\frac{1}{2266}$	$\frac{1}{2267}$	$\frac{1}{2268}$	$\frac{1}{2269}$	$\frac{1}{2270}$

should be mixed with 73 dwts. of pure gold to make a mass 23 carats fine? viz.:

Prussian Fred. d'or,	$21\frac{19}{32}$;	Denmark Ducat,	21;
Saxony August d'or,	$21\frac{18}{32}$;	Russian Imperial,	$21\frac{31}{32}$;
Baden, 10 Guilders,	$21\frac{19}{32}$;	Sicily, Onza,	$20\frac{20}{32}$;
Bruns'k, 10 Thalers,	$21\frac{16}{32}$;	Rome, Sequin,	$23\frac{29}{32}$;
Hesse, 10 Thalers,	$21\frac{11}{32}$;	do. Scudo,	20;
Belgic Sovereign,	22;	Sardinia, 40 lire,	$21\frac{18}{32}$.

Ans. 73 dwts. of the Roman Sequin, and $7\frac{5}{8}$ dwts. of each of the inferior rates.

Total alligation furnishes the rates of the several simples, with the rate and quantity of the composition, and requires the quantities of the simples.

RULE. Find the difference between the rates of the simples and the mean rate as before; then the sum of the differences is to the given quantity as each particular difference to its proper quantity.

Examples for the Slate.

1. A grocer has sugar at $12\frac{1}{2}$ cents, 15 cents, and 18 cents per lb., but wishing to compound 1 cwt. of these to be sold at 16 cents per lb., how much of each kind should be taken?

$$\text{Mean rate } 16 \begin{cases} 12.5 & & 2. \\ 15. & & 2. \\ 18. & 1+3.5= & 4.5 \end{cases}$$

$$\overline{8.5}$$

$$\text{Then, as } 8.5 : 112 :: \begin{cases} 2 & : 26\frac{6}{17} \text{ lbs. at } 12\frac{1}{2} \text{ cts.} \\ 2 & : 26\frac{6}{17} \text{ lbs. at } 15 \text{ “} \\ 4.5 & : 59\frac{5}{17} \text{ lbs. at } 18 \text{ “} \end{cases} \text{ Ans.}$$

2. A merchant has wines of $\$1\frac{1}{4}$, $\$1\frac{3}{4}$, and \$2 per gallon; but wishing to supply a particular customer with 50 gallons at $\$1\frac{1}{2}$, how much of each of the above kinds must be taken?

Ans. 30 gals. at $\$1\frac{1}{4}$, 10 gals. at $\$1\frac{3}{4}$, and at \$2.

2271	2272	2273	2274	2275	2276	2277	2278	2279	2280
$\frac{1}{2271}$	$\frac{1}{2272}$	$\frac{1}{2273}$	$\frac{1}{2274}$	$\frac{1}{2275}$	$\frac{1}{2276}$	$\frac{1}{2277}$	$\frac{1}{2278}$	$\frac{1}{2279}$	$\frac{1}{2280}$

3. How much gold of 15, 17, 19, and 23 carats fine must be taken to form a mass of 3 lbs. 20 carats fine?

Ans. 6 oz. of 15, 17, and 19 car., and 18 oz. of 23 car.

4. A crib, whose content is 100 bushels, being filled with wheat at \$1½, rye at \$1, and maize at 8 dimes per bushel, the mixture is worth \$1¼ per bushel; what quantity of each kind does it contain?

Ans. of wheat 58⅓ bu., of rye and maize each 20⅚.

SINGLE POSITION.

Position is a rule wherein supposed numbers are used to ascertain unknown ones, and is either Single or Double.

Single position is that in which the operation requires but one supposed number.

RULE. Assume any number and perform the operations with it that belong to the tenor of the question.

Then, As the result of the said operation
Is to the assumed number,
So is the result in the question
To the true position, or answer.

Demonstration. Let this proportion be represented thus, viz., $a : b : : c : d$. The results a and c are supposed to be deduced from the positions b and d by similar operations. Now these operations must consist, virtually, in adding to, or subtracting from the positions, similar multiples or parts of themselves, in which case a and c will be equimultiples or equal parts of b and d. But $2b : b : : 2d : d$, and $\frac{1}{2}b : b : : \frac{1}{2}d : d$ &c.; hence the rule is true. *q. e. d.*

Note. The questions usually given under the head of Position are easily solved by simple equations, or analysis; yet the rules of Position cannot be set aside simply because other modes of operation might effect the same end; for such a reason would exclude half the rules of arithmetic, since they are all abbreviations of the rule of counting one by one.

2281	2282	2283	2284	2285	2286	2287	2288	2289	2290
$\frac{1}{2281}$	$\frac{1}{2282}$	$\frac{1}{2283}$	$\frac{1}{2284}$	$\frac{1}{2285}$	$\frac{1}{2286}$	$\frac{1}{2287}$	$\frac{1}{2288}$	$\frac{1}{2289}$	$\frac{1}{2290}$

Examples for the Slate.

1. A person after spending $\frac{1}{3}$ and $\frac{1}{4}$ of his money, finds remaining $31\frac{1}{4}$\$; what sum had he at first?

suppose \$45

$\frac{1}{3}$= 15
$\frac{1}{4}$= 11.25

— 26.25

As 18.75 : \$45 : : \$31.25 : \$75 *Ans.*

2. What number is that which being multiplied by 8, and the product divided by 6, the quotient will be 40? *Ans.* 30.

3. A man being asked his age, said, if $\frac{1}{3}$ and $\frac{1}{4}$ of my age be multiplied by 4, and $\frac{1}{2}$ added to the product, the sum will be 102 years; Query the man's age? *Ans.* 36 yrs.

4. Four men divided their joint gain; A takes $\frac{1}{3}$, B $\frac{1}{4}$, C $\frac{1}{5}$, and D 65 dols., what was the whole sum? *Ans.* 300 dols.

5. A man receives for the amount of a sum at simple interest 5 years 728 dols.; the rate of interest being 6 per cent. what was the principal? *Ans.* \$560.

6. A schoolmaster being asked how many scholars he had, answered, "In addition to my present number, if I had half as many more, and $2\frac{1}{2}$, I should have 100," how many had he? *Ans.* 65.

7. Three pipes discharge 120 gallons of water from a cistern, severally as follows: the 1st in 20 minutes, the 2d in 40 minutes, and the 3d in 60 minutes; but if running at the same time, how much time will be required? *Ans.* $10\frac{10}{11}$ min.

If, in ex. 7, the 3d pipe supply water, in what time will the 1st and 2d discharge the 120 gallons? *Ans.* $17\frac{1}{7}$ min.

If the 2d be a supply pipe, what time will the 1st and 3d pipes require to discharge the water? *Ans.* 24 m.

2291	2292	2293	2294	2295	2296	2297	2298	2299	2300
$\frac{1}{2291}$	$\frac{1}{2292}$	$\frac{1}{2293}$	$\frac{1}{2294}$	$\frac{1}{2295}$	$\frac{1}{2296}$	$\frac{1}{2297}$	$\frac{1}{2298}$	$\frac{1}{2299}$	$\frac{1}{2300}$

If the 2d and 3d pipes supply water, in what time will the 1st discharge 120 gallons against the action of both? *Ans.* 120 min.

These questions, and similar ones, belong to the line of single fractions above.

8. A saves $\frac{1}{3}$ of his income; but B, who has the same salary, by living twice as fast as A, sinks $120 per annum; then what is the yearly sum? *Ans.* $360.

9. What sum, at 6 per cent., will amount to $592 in 8 years at simple interest? *Ans.* $400.

10. A captain being asked how many soldiers he had, replied, $\frac{1}{2}$ of them are in camp, $\frac{1}{3}$ in the trenches, $\frac{1}{8}$ in the hospital, and 4 in prison; of how many men did his company consist? *Ans.* 96 men.

DOUBLE POSITION.

Double Position requires two positions, or terms in the line of the prime series.

RULE. Assume two separate numbers, or positions, and perform the operations with each that belong to the tenor of the question; the differences between these results and the given result, are called the errors.

Multiply each error into the contrary position; if the errors are both too much, or both too little, divide the difference of the products by their difference; but if one is too much, and the other too little, divide the sum of the products by their sum; this result will be the answer.

Note. The difference between the true and first supposed number is to the difference between the true and second supposed number, as the first error is to the second; thus, in the 1st example, 60—45 : 48—45 : : $1\frac{1}{4}$: $\frac{1}{4}$; that is, 15 : 3 : : $\frac{5}{4}$: $\frac{1}{4}$, or 15 : 3 : : 5 : 1. It is only when this relation exists that the rule is applicable.

2301	2302	2303	2304	2305	2306	2307	2308	2309	2310
$\frac{1}{2301}$	$\frac{1}{2302}$	$\frac{1}{2303}$	$\frac{1}{2304}$	$\frac{1}{2305}$	$\frac{1}{2306}$	$\frac{1}{2307}$	$\frac{1}{2308}$	$\frac{1}{2309}$	$\frac{1}{2310}$

Examples for the Slate.

1. A son asking his father how old he was, received this answer: your age is now $\frac{1}{3}$ of mine, but 5 years ago your age was only $\frac{1}{4}$ of mine; then what are the ages of both?

1st supp. 60, $\frac{1}{3} = 20$ — 2d supp. 48, $\frac{1}{3} = 16$

$$\begin{array}{rrr} & 60 & 20 \\ & -5 & -5 \\ \hline & 55 & 15 \\ \frac{1}{4}= & 13\frac{3}{4} & -13\frac{3}{4} \\ \hline & & 1\frac{1}{4} \\ & & \times 48 \\ \hline & & 60 \end{array} \qquad \begin{array}{rrr} & 48 & 16 \\ & -5 & -5 \\ \hline & 43 & 11 \\ \frac{1}{4}= & 10\frac{3}{4} & -10\frac{3}{4} \\ \hline & & \frac{1}{4} \\ & & \times 60 \\ \hline & & 15 \end{array}$$

$1\frac{1}{4}$ errors both too little.

$$\text{Errors.}\left\{\begin{array}{r} 1\frac{1}{4} \\ -\frac{1}{4} \end{array}\right. \quad \left.\begin{array}{r} 60 \\ -15 \end{array}\right\}\text{Products.}$$

$$1\ )\ 45$$

Ans. 45 Father, $\frac{1}{3}=15$ Son.

$$\begin{array}{r} 45 \\ -5 \\ \hline 40 \end{array}$$

40, $\frac{1}{4}=10+5=15$ Proof.

2. What number is that which, being multiplied by 6, the product increased by 18, and that sum divided by 9, will result in 20? *Ans.* 27.

3. The age of A is 18 years, of B as much and the half of C's, of C as much as that of both the former; how much is the age of each?

Ans. A's 18, B's 54, and C's 72 years.

4. The head of a fish is 8 inches long, the tail is as long as the head and half the body, the body is as long as the head and tail; required the length of the fish?

Ans. 5 feet, 4 inches.

5. Paid for 18 yards of broadcloth and lining $34.8, the cloth was $5 per yard, and the lining 40 cents; what was the quantity of each? *Ans.* cloth 6, and lining 12 yards.

2311	2312	2313	2314	2315	2316	2317	2318	2319	2320
$\frac{1}{2311}$	$\frac{1}{2312}$	$\frac{1}{2313}$	$\frac{1}{2314}$	$\frac{1}{2315}$	$\frac{1}{2316}$	$\frac{1}{2317}$	$\frac{1}{2318}$	$\frac{1}{2319}$	$\frac{1}{2320}$

6. A laborer being employed 20 days, at $1 per day if he worked, otherwise to forfeit per day .75$, now at settlement he was paid $11.25; how many days of the time did he work? *Ans.* 15 days.

7. A saddle, worth $50 on the back of Swift, makes his value double that of Chestnut; but the saddle set on Chestnut makes him worth $\frac{4}{5}$ as much as Swift; what is the value of each horse?

Ans. Swift 250 dollars, Chestnut 150 dollars.

8. A and B began to play together with equal sums of money; A first won 20 guineas, but afterward lost back $\frac{2}{3}$ of what he then had; after which, B had 4 times as much as A; what sum did each begin with?

Ans. 100 guineas.

9. When the beautiful Caroline first became mine,
Our years were proportioned as three is to nine;
But of these, since our union, three-fourths of a score
Having passed, the proportion* is now two to four,
Say, can William by science of numbers define,
At her marriage the age of my sweet Caroline?

Ans. Sir, the age of your Caroline plainly is seen,
At the barely blown bloom of the youthful fifteen.

PERMUTATIONS.

This is a rule for showing the variations or changes of position which any number of things can admit; and these changes may be of the whole, taken one by one or singly; or two by two, called binary; or three by three, called ternary, &c., until all the things are taken at every change.

---

* In ordinary discourse, proportion and ratio are used indifferently; but mathematical strictness forbids this license. The meaning therefore is, that the ratio of her age to mine is as that of 2 to 4.

2321	2322	2323	2324	2325	2326	2327	2328	2329	2330
$\frac{1}{2321}$	$\frac{1}{2322}$	$\frac{1}{2323}$	$\frac{1}{2324}$	$\frac{1}{2325}$	$\frac{1}{2326}$	$\frac{1}{2327}$	$\frac{1}{2328}$	$\frac{1}{2329}$	$\frac{1}{2330}$

RULE. The continual product of the terms of the prime series, beginning at the number of things, downward, to as many terms as there are things taken at once, will be the number of changes.

Illustration. If 8 things be taken singly, there will be 8 changes; if 2 by 2, the changes will be 8×7=56; by ternary changes, they make 8×7×6=336, &c.; and if all be taken, it will be the continual product of the series down to unity.

Examples for the Slate.

1. How many different positions may 7 persons assume at a table, the permutations being total? *Ans.* 5040.
2. How many tickets are in that lottery which admits of 30 ternary permutations of the numbers? *Ans.* 24360.
3. How many total changes may be made of the words in the following verse: *Audi, vide, tace, Si tu vis vivere pace?* *Ans.* 40320.
4. How many words of 5 letters may be made out of 24, admitting that consonants alone will make a word? *Ans.* 5100480.

Binary Permutations of Six Letters, a, b, c, d, e, f.

ab, ac, ad, ae, af; ba, bc, bd, be, bf; ca, cb, cd, ce, cf; da, db, dc, de, df; ea, eb, ec, ed, ef; fa, fb, fc, fd, fe: 30=6×5, as per Rule.

COMBINATIONS.

The last rule respects only the position of things; for there the same things are taken repeatedly, but in a different order. In this rule the associations must always be new, and at least of two things taken together. Of the whole at once there can be only one combination. The combinations are binary, ternary, quaternary, &c.,

2331	2332	2333	2334	2335	2336	2337	2338	2339	2340
$\frac{1}{2331}$	$\frac{1}{2332}$	$\frac{1}{2333}$	$\frac{1}{2334}$	$\frac{1}{2335}$	$\frac{1}{2336}$	$\frac{1}{2337}$	$\frac{1}{2338}$	$\frac{1}{2339}$	$\frac{1}{2340}$

as the things are taken two by two, three by three, four by four, &c.

RULE. Divide the result according to the last rule, by the continual product of the series, from unity to as many terms as there are particulars in one combination; the quotient will be what you desire.

Illustration. If 8 things be taken 2 by 2, it will be $(8\times7)\div(1\times2)=(56\div2)=28$; or, 3 by 3, it makes $(8\times7\times6)\div(1\times2\times3)=(336\div6)=56$; &c.

Examples for the Slate.

1. How many tickets are in that lottery, the numbers of which admit of 30 ternary combinations? *Ans.* 19600.

2. The tickets of another lottery admit 60 ternary combinations, what is the number of them? *Ans.* 34220.

3. How many files of 10 men may be elected from a company consisting of 100 men? *Ans.* 17310309456440.

Binary Combinations of Six Letters, a, b, c, d, e, f.

ab, ac, ad, ae, af; bc, bd, be, bf; cd, ce, cf; de, df; ef: $15=(6\times5)\div(1\times2)$.

Ternary Combinations of Six Letters, a, b, c, d, e, f.

abc, abd, abe, abf; acd, ace, acf; ade, adf; aef; bcd, bce, bcf; bde, bdf; bef; cde, cdf; cef; def: $20=(6\times6\times4)\div(1\times2\times3)$, as per Rule.

MENSURATION OF ARTIFICERS' WORK.

When two dimensions are given, their product will be the area, or superficial content. In solids, the area is multiplied into the depth or thickness.

RULE. Feet×feet=feet, feet×inches=inches, feet×sec-

2341	2342	2343	2344	2345	2346	2347	2348	2349	2350
$\frac{1}{2341}$	$\frac{1}{2342}$	$\frac{1}{2343}$	$\frac{1}{2344}$	$\frac{1}{2345}$	$\frac{1}{2346}$	$\frac{1}{2347}$	$\frac{1}{2348}$	$\frac{1}{2349}$	$\frac{1}{2350}$

onds=seconds, inches×inches=seconds, inches×seconds =thirds, seconds×seconds=fourths; but the smaller denominations are rejected in the dimensions.

Note. The rules of Compound Multiplication and Practice explain all that is necessary in these operations.

Brick work is computed by the rod of $16\frac{1}{2}$ feet, whose square is $272\frac{1}{4}$; this is called superficial measure, but the standard thickness of $1\frac{1}{2}$ bricks is understood.

Glazing is computed by the square foot, masonry by the cubic foot; painting, plastering, paving, &c., by the square yard; flooring, partitioning, roofing, &c., by the square of 100 feet; viz., 10^2.

Note. In the following examples, the rules for taking the dimensions are generally included.

Examples for the Slate.

1. What is the content of a wall in rods, the compass being 68 ft. $0\frac{3}{4}$ in., taken $\frac{1}{2}$ within and $\frac{1}{2}$ on the outside of the building; the height 15 ft., and the thickness 2 bricks?

	ft.	in.
	68	$0\frac{3}{4}$
		$3\times5=15$ ft.
	204	$2\frac{1}{4}$
		5
	1020	$11\frac{1}{4}$ at given thickness.
$\frac{4}{3}=\frac{3}{3}+\frac{1}{3}$	340	$3\frac{3}{4}$ for the $\frac{1}{2}$ brick more.
$272\frac{1}{4}$)	1361	3 at standard thickness.
4	4	
1089)	5445	(5 rods. *Ans.*
	5445	

2. A triangular gable (which contains the product of $\frac{1}{2}$ the perpendicular height into the base) is raised $17\frac{1}{2}$ ft. on an end wall 24 ft. 9 in. long; the thickness being 2 bricks, what will the whole measure, in square yds.?

Ans. $32.08\frac{1}{3}$ yards.

2351	2352	2353	2354	2355	2356	2357	2358	2359	2360
$\frac{1}{2351}$	$\frac{1}{2352}$	$\frac{1}{2353}$	$\frac{1}{2354}$	$\frac{1}{2355}$	$\frac{1}{2356}$	$\frac{1}{2357}$	$\frac{1}{2358}$	$\frac{1}{2359}$	$\frac{1}{2360}$

3. The end wall of a house is 28 ft. 10 in. long, and 55 ft. 8 in. high, to the eaves; of which height, 20 ft. is $2\frac{1}{2}$ bricks, 20 ft. is 2 bricks, and $15\frac{2}{3}$ ft. is $1\frac{1}{2}$ bricks; reduced to the standard, what is the content?

Ans. $8\frac{134}{9801}$ rods.

4. How many cubic feet of mason work does that wall contain, whose length is $53\frac{1}{2}$ ft., height $12\frac{1}{4}$, and thickness 2 feet? *Ans.* $1310\frac{3}{4}$ feet.

5. In a chimney piece, suppose the length of the mantle and slab each $4\frac{1}{2}$ ft., breadth of both together 3 feet 2 in., length of each jamb $4\frac{1}{3}$ ft., breadth of these together $1\frac{3}{4}$ feet; what is the superficial content?

Ans. 21 ft. 10 in.

6. What is the content in squares of 100 ft. of a floor $48\frac{1}{2}$ ft. long and $22\frac{1}{4}$ ft. broad?

Ans. $10.79\frac{1}{8}$ squares.

7. An arched cellar is $16\frac{1}{2}$ feet long, and by passing the line over the surface of the arch, the width is ascertained to be 22 feet 8 in.; how many square yards of centering does it require? *Ans.* 41 yds. 5 ft.

8. By passing the line close over the tread and riser, from top to bottom, a case of stairs is found to be 62 feet 8 in. long, and round the front and returns of a step measures 5 feet 6 in. for the breadth; required the content? *Ans.* $38\frac{8}{27}$ sq. yds.

9. Passing the line, from the top of the newel post, over the whole length of the handrail, a balustrade is found to be $48\frac{1}{2}$ feet long, and twice the length of the baluster on the landing with the girt of the handrail, taken for the breadth, is $6\frac{1}{2}$ feet; required the content?

Ans. $35\frac{1}{36}$ yards.

10. The girt of the architrave and jamb of a door frame is $8\frac{1}{6}$ feet for the breadth; the length being the sum of the lengths of the two jambs and lintel is $17\frac{1}{2}$ feet, what does it measure? *Ans.* $15\frac{71}{108}$ yards.

11. A door 7 ft. by $3\frac{1}{2}$ feet, is panneled on both sides, for which it is reckoned double; how many yards of work will it measure? *Ans.* 5 yards 4 ft.

12. The shutters for 3 windows measure 7 feet by $3\frac{1}{2}$

2361	2362	2363	2364	2365	2366	2367	2368	2369	2370
$\frac{1}{2361}$	$\frac{1}{2362}$	$\frac{1}{2363}$	$\frac{1}{2364}$	$\frac{1}{2365}$	$\frac{1}{2366}$	$\frac{1}{2367}$	$\frac{1}{2368}$	$\frac{1}{2369}$	$\frac{1}{2370}$

feet and are panneled only on one side, for which the area and half is taken; how many square yards of work do they contain? *Ans.* $12\frac{1}{4}$ yards.

13. How many squares of 100 feet does the roof measure, when the house is $44\frac{1}{2}$ ft. long and $18\frac{1}{4}$ feet broad; allowing the breadth over the ridge to be 3 halves of the flat, which is nearly true, when the roof forms a right angle at the top? *Ans.* $12.18\frac{3}{16}$ squares.

14. How many yards of rendering (plastering on walls) are in that room, the compass of which is 55 feet 2 in. and height $9\frac{1}{4}$ feet? *Ans.* 56 yds. 6 ft. $3\frac{1}{2}$ in.

15. How many yards of ceiling (plastering on laths) are in a room $14\frac{1}{2}$ feet long, and $11\frac{3}{4}$ feet broad?
Ans. 18 yds. $8\frac{3}{8}$ feet.

16. The length of a room is 20 feet, breadth $14\frac{1}{2}$ feet, and height 10 ft. 4 inches, deducting a fire-place 4 feet by $4\frac{1}{3}$ ft. and 2 windows each 6 feet by $3\frac{1}{6}$ feet; how many yards of painting does it contain? *Ans.* $105\frac{8}{27}$ yds.

17. An oval fan-light is $14\frac{1}{2}$ ft. long, and $4\frac{3}{4}$ ft. wide; how many feet of glazing will it measure, allowing nothing for the elliptical form on account of waste in cutting? *Ans.* 68 ft. $10\frac{1}{2}$ in.

18. The diameter of a semicircular window is 7 ft. 9 in., and its height 3 ft. $10\frac{1}{2}$ in., what will the glazing amount to, at 25 cents per square foot?
Ans. $7\frac{65}{128}$ dols.

Note. In this example, because the window is a semicircle, the height is half the diameter; and the product of the diameter and height is taken for the area, without deduction.

19. How much will the paving of a yard amount to, dimensions 63 ft. by 45; a walk $5\frac{1}{4}$ feet wide, running the whole length, being flagged, at $1 per yd.; and the rest paved with pebbles at 75 cts. per yard?
Ans. $245.4375.

20. What will the covering and guttering of a roof with lead amount to, at $6\frac{1}{4}$ cts. per lb., length 40 feet, breadth over the ridge 32 ft., the guttering 56 ft. by 2 ft., the lead being $8\frac{1}{2}$ lbs. to the square foot?
Ans. 739.5 dols.

2371	2372	2373	2374	2375	2376	2377	2378	2379	2380
$\frac{1}{2371}$	$\frac{1}{2372}$	$\frac{1}{2373}$	$\frac{1}{2374}$	$\frac{1}{2375}$	$\frac{1}{2376}$	$\frac{1}{2377}$	$\frac{1}{2378}$	$\frac{1}{2379}$	$\frac{1}{2380}$

MENSURATION OF TIMBER.

Case 1. To find the superficial content of a board or plank.

RULE. If the board be of equal breadth, the product of the length and breadth will be the answer.

Note. If the board be tapering, take half the sum of the breadths at both ends, or the breadth in the middle for the mean breadth.

Examples for the Slate.

1. What is the value of a plank whose length is 8 ft. 6 in. and mean breadth 1 ft. 3 in. at 5 cts. a foot?
Ans. 53$\frac{1}{8}$ cts.

2. What is the value of a plank whose length is 11 ft. 6 in., and breadth 13 in., at 10 cts a foot?
Ans. $1.25 nearly.

3. How many square feet will 6 boards measure, 3 of which are 16$\frac{1}{2}$ ft. long, and 14 in. broad each, and 3 others 18 ft. 4 in. long, and their united breadth 3 ft. 9 in.?
Ans. 126 ft. 9 in.

4. What is the value of 5 oaken planks, at 4 cts. per foot, their common length 19$\frac{1}{2}$ ft., the breadth of 3 taken together being 33$\frac{1}{4}$ in., and of the other two, each at the broad end 18 in., and 10$\frac{1}{2}$ in. at the narrow end?
Ans. 4.01\frac{3}{8}$.

Case 2. To find the solid content of squared and round timber.

RULE. Multiply the mean breadth by the mean thickness, and that product by the length for the answer.

Note 1. In round timber the quarter girt, i. e. $\frac{1}{4}$ of the mean circumference, will be the mean breadth and thickness.

2. If the tree taper regularly, take the mean breadth and thickness in the middle, or take half the sum of the dimensions at both ends, for the mean dimensions.

3. If the piece be thick in some parts and small in others, take several different dimensions, and divide their sum by the number of them for the mean dimensions.

2381	2382	2383	2384	2385	2386	2387	2388	2389	2390
$\frac{1}{2381}$	$\frac{1}{2382}$	$\frac{1}{2383}$	$\frac{1}{2384}$	$\frac{1}{2385}$	$\frac{1}{2386}$	$\frac{1}{2387}$	$\frac{1}{2388}$	$\frac{1}{2389}$	$\frac{1}{2390}$

Examples for the Slate.

1. The length of a piece of timber is 18 ft. 6 in., the breadths at the greater and less ends 1 ft. 6 in. and 1 ft. 3 in., and the thickness at the greater and less ends 1 ft. 3 in. and 1 ft., required the solid content?

Ans. 28 ft. 7+in.

2. What does the piece of timber contain whose length is $24\frac{1}{2}$ ft. and mean breadth and thickness each 1.04 ft. *Ans.* $26\frac{1}{2}$—feet.

3. Required the content of a piece of timber, whose length is 20.38 ft., breadth and thickness at the greater end $19\frac{1}{8}$ in. each, and at the less $9\frac{7}{8}$ in. each?

Ans. 29.7562+ft.

RULES FOR SUMMING THE SERIES OF TRIANGLES, SQUARES, RECTANGLES, AND CUBES.

See the formation of these series in the Introduction and in Simple Multiplication.

1. *Of the Triangular series.*

The terms of this series are formed from the consecutive additions of the terms of the prime series: they are 1, 3, 6, 10, 15, 21; we stop at 6 terms, and let n represent the number.

The *formula* for this is, $\frac{1}{6}(n+2)\times(n+1)\times n=s$.

And substituting 6 for n, it will be

$\frac{1}{6}(6+2)\times(6+1)\times6=8\times7=56$, the sum.

Again, let the terms be 1, 3, 6, 10, 15, 21, 28, 36; eight in number, in which $n=8$.

This will be $\frac{1}{6}(8+2)\times(8+1)\times8=120$, the sum.

2. *Of the series of Squares.*

The terms of this series are formed from the consecutive additions of the terms of the odd series: they are 1,

2391	2392	2393	2394	2395	2396	2397	2398	2399	2400
$\frac{1}{2391}$	$\frac{1}{2392}$	$\frac{1}{2393}$	$\frac{1}{2394}$	$\frac{1}{2395}$	$\frac{1}{2396}$	$\frac{1}{2397}$	$\frac{1}{2398}$	$\frac{1}{2399}$	$\frac{1}{2400}$

4, 9, 16, 25, 36 ; we stop at 6 terms, and let n represent the number.

The *formula* for this is, $\frac{1}{6}(n+1)\times(2n+1)\times n=s$.

And substituting 6 for n it will be

$\frac{1}{6}(6+1)\times(12+1)\times 6=7\times 13=91$, the sum.

Again, let the terms be 1, 4, 9, 16, 25, 36, 49, 64 ; eight in number, in which $n=8$.

This will be $\frac{1}{6}(8+1)\times(16+1)\times 8=204$, the sum.

3. *Of the series of Rectangles.*

The terms of this series are formed as in simple multiplication ; viz., 1×2, 2×3, 3×4, 4×5, &c., or 1×3, 2×4, 3×5, 4×6, 5×7, &c. And it is evident that the number of series of this class may be infinite ; in all of which the dimensions are unequal.

Let the terms be 2, 6, 12, 20, 30, 42, stopping at 6 terms; and let n be the number, and let m be the least term less 1. The *formula* for this will be $\frac{1}{6}(2n+1+3m)\times(n+1)\times n=s$.

Then substituting 6 for n, and 2—1 for m, it will be

$\frac{1}{6}(2\times 6+1+3\times 1)\times(6+1)\times 6=16\times 7=112$, the sum.

Again : let the terms be 2, 6, 12, 20, 30, 42, 56, 72; eight in number, in which $n=8$, and $m=2-1$, as before.

This will be $\frac{1}{6}(2\times 8+1+3\times 1)\times(8+1)\times 8=240$, sum.

4. *Of the series of Cubes.*

The sum of the Cubes of the terms of the prime series, 1, 2, 3, &c., is equal to the square of the sum of the roots, that is, the square of the sum of the same series ; thus, $1^3+2^3=(1+2)^2=3^2=9$; $1^3+2^3+3^3=(1+2+3)^2=6^2=36$; $1^3+2^3+3^3+4^3=(1+2+3+4)^2=10^2=100$.

The sum of the arithmetical series referred to above, is equal to the sum of the extremes into $\frac{1}{2}$ the number of terms ; thus $1+2+3+4+$ &c. to $20=1+20\times\frac{20}{2}=21\times 10=210$. And the square of all the Cubes from 1 to 20 inclusive will be $210^2=44100$.

What is the sum of the Cubes of the terms of the prime series to 40 terms ? *Ans.* 672400.

2401	2402	2403	2404	2405	2406	2407	2408	2409	2410
$\frac{1}{2401}$	$\frac{1}{2402}$	$\frac{1}{2403}$	$\frac{1}{2404}$	$\frac{1}{2405}$	$\frac{1}{2406}$	$\frac{1}{2407}$	$\frac{1}{2408}$	$\frac{1}{2409}$	$\frac{1}{2410}$

GLOSSARY OF TERMS

USED IN THE PRECEDING WORK, TECHNICALLY DEFINED, AND ARRANGED IN THE ORDER OF THEIR INITIAL LETTERS.

A.

Acceptor. He who assumes the payment of a bill, by writing upon it his name after the word accepted.

Addition. The collecting of many sums into one.

Addition Simple. The collecting of many terms of the prime series into one.

Addition of Decimals. The collecting of numbers of several *ranks*, positive, or negative, or both, into one sum.

Addition of Mercantile Numbers. The collecting into one sum of numbers, whose units are of several grades or measures, and therefore require to be reduced to the same grade or measure.

Addition of Vulgar Fractions. The collecting of parts into one sum; but the parts must have a common denominator, or be reduced to such, before they can be so collected.

Aggregate. The sum, the amount, the result of addition.

Agio. The difference between banco and current money in the Netherlands.

Algorithm. The addition, subtraction, and all other operations of numbers.

Alligation. The joining of alternate simples to each other, on either side of a mean rate, and giving to each simple the difference between the mean rate and the simple, with which each is joined.

Alternate. The change of one for another;—in alligation, on either side of the mean rate;—in proportions, the ratio of the first and second, or first and third terms.

Amount. The aggregate, the sum of principal and interest, also of the annuity and its interest.

Analysis (Resolution). The method of separating general propositions into their simple elements.

Angular Measure. The measure applied to circles, of which a degree is the unit.

Annuities. Periodical payments arising from houses, lands, stocks, &c.

Antecedent. The leading term of a ratio, or couplet; as 3 in the ratio of 3 to 6.

Apothecaries' Weights. The small measures of gravitation applied only to medicines.

Arithmetic. The science of the numerical series.

2411	2412	2413	2414	2415	2416	2417	2418	2419	2420
$\frac{1}{2411}$	$\frac{1}{2412}$	$\frac{1}{2413}$	$\frac{1}{2414}$	$\frac{1}{2415}$	$\frac{1}{2416}$	$\frac{1}{2417}$	$\frac{1}{2418}$	$\frac{1}{2419}$	$\frac{1}{2420}$

Arithmetical Progression. The term applied to that class of series which are formed by equal additions and subtractions.

Artificers' Work. The work of mechanics, builders, &c.

Avoirdupois Weights. The measures of gravitation applied to coarse metals, groceries, and other articles.

B.

Bank Discount. The interest retained by banks, for the money they lend.

Barter. The exchange of commodities.

Bills of Exchange. Orders for the payment of money at sight, at usance, or at times specified therein.

Billions. The third period of multiples in the decimal scale, including six ranks, viz.: 1.000,000.000,000; 22.000,000.000,000; &c., to 999,999.000,000.000,000.

Billionths. The second period of parts in the decimal scale, including six ranks, viz.: .000,000.000,001; .000,000.000,021; &c., .000,000.999,999.

Binary. In combinations, two by two.

Biquadrate. The second power of a square; the fourth power of a root.

Breadth. One of the two dimensions of a surface.

Brokerage. The commission of a broker, usually a rate per cent., an allowance on the 100.

C.

Carat. 1-24th part of a mass of gold or silver.

Centum. One hundred; per cent. by the 100.

Circle Measure. See angular measure.

Circular Remittances. The sending of money from country to country, for the benefit to be derived from exchanges.

Circulating Decimals. Figures quoted again and again in division, renewing the same ranks without end; because the dividend, however often multiplied by 10, never becomes a multiple of the divisor. (See Note, Page 158.)

Cloth Measure. The measure of extension, which is divided into yards, quarters, nails, inches.

Coins. Moneys, small pieces of gold, silver or copper, stamped with the image and superscription of sovereignty; and authorized by law as a medium of exchange.

Combinations. The connecting of things two by two, three by three, &c., without repeating the same connection.

Commissions. Per centum allowances made for transacting the business of others.

Common Difference. The number used in forming simple series, by subtraction; the constant difference of the terms in arithmetical progressions.

Common Increase. The number constantly added in forming a simple series, beginning at the least term.

2421	2422	2423	2424	2425	2426	2427	2428	2429	2430
$\frac{1}{2421}$	$\frac{1}{2422}$	$\frac{1}{2423}$	$\frac{1}{2424}$	$\frac{1}{2425}$	$\frac{1}{2426}$	$\frac{1}{2427}$	$\frac{1}{2428}$	$\frac{1}{2429}$	$\frac{1}{2430}$

Common Measure. A measure which will exactly divide two or more numbers.

Common Multiple. A number which is a multiple of two or more numbers, and is therefore a common multiple.

Complement. That which completes or fills up; as, a single part and its complement make a unit.

Composite Numbers. All multiples of two, three, or any number above one, are called composite.

Composition (Synthesis). The method of reasoning from the simplest elements to the more complex combinations.

Compound Fraction. A part of a part; as $\frac{1}{2}$ of $\frac{1}{3}$, $\frac{3}{4}$ of $\frac{5}{6}$, &c.

Compound Ratio. The product of two or more ratios; viz., the product of their antecedents for an antecedent; and the product of their consequents for a consequent.

Compound Interest. The interest of the principal and former interest.

Continued Fractions. Such fractions as have a mixed number for one of the terms.

Consequent. The latter term of a couplet or ratio; as 6 in the ratio of 3 to 6.

Couplet. The terms of a ratio, the antecedent and consequent.

Course of Exchange. The commercial price of Foreign bills.

Cross. Erect $+$, the sign of addition, read *plus*, or more; oblique $\times$, the sign of multiplication, read *times*, *parts of*, or *multiplied unto*.

Cube. The third power of a number, the unit of measure in all solids.

Cubic. In the form of a cube, having three equal dimensions.

Cube Cubed. The third power of a cube; the ninth power of the root.

Cube Squared. The second power of a cube, the sixth power of the root.

Cylinder. A round solid, whose opposite ends are parallel and equal circles.

D.

Data. The given parts in every rule or question.

Date. The given year, month and day.

Decimal. Belonging to ten; tenth multiples, or tenth parts, the ranks of the decimal notation.

Decimal Fraction. A fraction the denominator of which is 10, 100, 1000, or some term of this series.

Denominators. The ordinal numbers, viz., first or prime, second, third, fourth, &c.; the lower term in fractions, answering to the question, what part?

Difference. The excess of one number above another, the remainder.

Digits. The ten Arabic figures, from *digiti*, fingers.

Direct Proportion. That in which the consequents are both less, or both greater, than the antecedents

2431	2432	2433	2434	2435	2436	2437	2438	2439	2440
$\frac{1}{2431}$	$\frac{1}{2432}$	$\frac{1}{2433}$	$\frac{1}{2434}$	$\frac{1}{2435}$	$\frac{1}{2436}$	$\frac{1}{2437}$	$\frac{1}{2438}$	$\frac{1}{2439}$	$\frac{1}{2440}$

Discount. The sum retained by the lender for the interest of his money.

Dividend. The number to be divided, one of the given terms in division.

Dividual. A part of the dividend, from which one figure results to the quotient.

Division, Simple. That in which the divisor and dividend are terms of the prime series.

Division of Decimals. That in which the dividend and divisor may consist of ranks on either side of the units' rank, and the negative ranks of the divisor and quotient together will equal those of the dividend.

Division of Mercantile Numbers. That in which remainders of higher grades must be reduced to the lower, and added to the lower given grades, in order to carry out the division.

Division of Vulgar Fractions. That in which the terms of the divisor must be inverted, and used as multipliers, unless when they are exact measures of the terms of the dividend respectively.

Division, Long. That in which the successive multiplications and divisions are exhibited.

Division, Short. That in which the quotient only appears; and the multiplications and subtractions are mental.

Division of Joint Stocks. A proportional division, depending upon the ratios of the stocks only, or upon the ratios composed of the ratios of the stocks and times.

Divisor. The measure applied to the dividend, one of the given terms in division.

Drawer. One of the parties concerned in a bill of exchange.

Dry Measure. The capacious measure of which the divisions are the bushel, peck, quart, pint.

Duodecimal. Belonging to 12, a division of long measure into 4ths, 3ds, 2ds, inches, feet, in which the scale is 12, or 12ths. It has heretofore been practised in cross multiplication, a slow method, now nearly obsolete.

Duration. A general property of things measured by the motion of bodies, and numbered by years, months, weeks, days, hours, minutes, seconds.

Note. Unless we call duration a property of *things*, we must leave the physical world, and employ our arithmetic upon the unknown, uncreated nonentity, *extra flammantia mœnia mundi.*

E.

Equation of Dates. A rule for finding a mean date between the dates of two bills of parcels, such that the sum of the two bills is to the whole line of intervening time, as each bill is to its portion of the time measured from the opposite point.

Equation of Payments. The finding of a mean time for discharging several payments proportional to the units of each, both in money and time, multiplied together.

2441	2442	2443	2444	2445	2446	2447	2448	2449	2450
$\frac{1}{2441}$	$\frac{1}{2442}$	$\frac{1}{2443}$	$\frac{1}{2444}$	$\frac{1}{2445}$	$\frac{1}{2446}$	$\frac{1}{2447}$	$\frac{1}{2448}$	$\frac{1}{2449}$	$\frac{1}{2450}$

Equal Parts. The same parts of two or more numbers; as, their halves, thirds, fourths, &c.

Equimultiples. The same multiple of two or more numbers; for since every number has a series of multiples, of which itself is the first, its double is the second, its triple is the third, &c., the same multiple of two numbers will be their equimultiples, though any multiple of the greater will be greater than the same multiple of the less.

Erect Cross. See Cross.

Even Numbers. 2, and its multiples in the prime series, are all even numbers.

Evolution. A rolling out, that is, of the root; the extracting of roots.

Exchange. The comparison of the moneys of different countries, made here chiefly with the dollar.

Note. Under the head of Exchange a great variety of gold and silver coins are introduced, stating the value of each in Federal Money; and in Alligation the ounce value of many is given, with the fineness of each in carats, and 32d parts of a carat.

Exponents. The indices of powers, which are marked with the positive or negative sign +, or —, according as the powers occupy ranks on the left or right of the units' rank.

Extension. One of the general properties of things which are measured and numbered.

Note. The simplest form of extension is length only; the second form consists of length and breadth, and the third form consists of length, breadth, and thickness; and when any measure is applied to these dimensions, as a foot, a yard, &c., the divisions are numbered, each division being equal to the measuring unit.

F.

Factors. The multiplicand and multiplier.

Federal Money. The money of the United States, of which the divisions are decimal, and the unit $1.

Formula. A little form consisting *here* of initial letters and operative signs, and exhibiting rules and modes of operation. The rules, however, are always expressed plainly in words, in addition to the formulæ.

Fourth Proportional. The fourth term of a set of four proportionals; it may be found by multiplying the 3d term by the quotient of the 2d and 1st; or multiplying the 2d by the quotient of the 3d and 1st; or dividing the product of the 2d and 3d by the 1st.

Fractions. Parts of numbers, distinguished into single parts, plural parts, proper fractions, improper, simple, compound, mixed and continued fractions.

G.

Geometrical Progression. The class of series which increase and decrease by a constant multiplier, or ratio. The series increase or decrease as the ratio may be greater or less than a unit.

Grades. A term used here to designate units of measure of the same kind, but of different magnitudes: Mercantile numbers.

2451	2452	2453	2454	2455	2456	2457	2458	2459	2460
$\frac{1}{2451}$	$\frac{1}{2452}$	$\frac{1}{2453}$	$\frac{1}{2454}$	$\frac{1}{2455}$	$\frac{1}{2456}$	$\frac{1}{2457}$	$\frac{1}{2458}$	$\frac{1}{2459}$	$\frac{1}{2460}$

Gravitation. One of the general properties of things which are measured and numbered: This property is measured by three classes of weights, viz., Troy, Avoirdupois, and Apothecaries' Weights.

Greatest Term. The last of an increasing series, and first of a decreasing one.

H.

Hundreds. Units of the 3d positive rank; as 100, 200, 300, &c., to 900; the 3d rank of every half period in the decimal scale.

Hundredths. Units of the 1st and 2d negative ranks in the decimal scale; as .01, .02, &c. to .99.

Half. The second part of a number; or used instead of second.

I.

Improper Fractions. Those in which the numerators are equal to, or greater than the denominators.

Index, Pl. *Indices.* The exponents of powers of numbers.

Note. The indices of positive ranks, are marked with the sign plus +; those of negative ranks are marked with the sign minus —.

Indorser. He who makes himself responsible for the payment of a note, by writing his name upon the back of it.

Induction (in Arithmetic). That form of practice which leads into the laws of the several series.

Infinite. Without limit as to number of terms; this is the only infinity known in the mathematics; the infinite divisibility of a finite quantity, the terms of an interminable series.

Ingredients. Simples taken to form a composition.

Insurance. A premium paid for security against loss of property exposed to risk.

Interest, Simple. The premium allowed for money lent, or for the principal only.

Interest, Compound. The premium required for the use of a principal sum, and its former interest.

Inverse Proportion. The contrast of two equal ratios; not the inversion of both couplets, as in Eu. b. 5. p. B, but only of one of them. It is applied in the action and reaction of forces, and may be exemplified in any of the mechanic powers.

Involution. A continual multiplication of a number by itself; in which the number itself is the first power, and every product is a power: the powers are marked by the terms of the prime series as indices; and these indices are marked with the signs + plus, or — minus to denote their positive or negative ranks.

L.

Lemma. A proposition assumed, in order to introduce more clearly the doctrines which are to follow.

Least Term. The first term of an increasing series, and the last of a decreasing one.

2461	2462	2463	2464	2465	2466	2467	2468	2469	2470
$\frac{1}{2461}$	$\frac{1}{2462}$	$\frac{1}{2463}$	$\frac{1}{2464}$	$\frac{1}{2465}$	$\frac{1}{2466}$	$\frac{1}{2467}$	$\frac{1}{2468}$	$\frac{1}{2469}$	$\frac{1}{2470}$

Length. The simplest form of extension, a line. The measures applied to a line, are the inch, foot, yard, pole, rod or perch, furlong, mile, league, degree; and these are called long measures.

Liquid Measures. The measures of capacity applied to fluids. These are the gallon, quart, pint, gill. Barrels, hogsheads, puncheons, &c., are the names of vessels, not of measures.

M.

Magnitude. Extension without measure, and of course without number.

Means. The terms which go between the extremes in a series, or in a set of proportionals: Means is the present participle of *meo*, Latin, and is singular though ending in *s*. It is usual, however, in the Mathematics to say, *one mean, term* being understood.

Mean Proportional Term. The middle term in an odd set of proportionals; as a unit between 10 and $\frac{1}{10}$, 100 and $\frac{1}{100}$, &c.

Measure. A standard of comparison, to which magnitudes of all kinds are brought, in order to have them divided and numbered; for whatever is measured is divided into units equal to the measure; and these units are called one, two, three, &c., that is, they are numbered.

Note. Measures are of three kinds, viz.. of extension, gravitation, and duration.

Medial. In Alligation, a term applied to the case in which the rates and quantities of the simple ingredients are *given*, to find the rate of the composition, or mean rate.

Mercantile Numbers. Concrete numbers, or such as are applied to money, merchandise, labor, time, or any species of property, which may be measured and numbered.

Millions. The second period of multiples in the decimal scale, including six ranks, viz.: 1.000,000, 32.000,000, &c., to .999,999.000,000.

Millionths. The first period of parts in the decimal scale, including six ranks, viz.: .000,001, .000,032, &c., to .999,999.

Minuend. To be lessened, the greater given term in subtraction.

Mixed Numbers. Such numbers as have two measuring units, one of which is a single part of the other; as in $3\frac{3}{4}$, where the measuring units are 1, and $\frac{1}{4}$ of 1; also in 5.25, where the measuring units are 1, and .01, or $\frac{1}{100}$ of 1.

Money. See Coins.

Multiple, Multiplex, Manifold. Every number has a series of multiples, of which itself is the first, its double is the second, its triple is the third, &c.

Multiplier. The lower factor, a given term in multiplication, showing how many folds of the multiplicand make up the product.

Multiple, Common. A number which has two or more exact measures; thus, 6 is the 2d multiple of 3, and the 3d multiple of 2.

Multiplicand. To be multiplied, a given term in multiplication, the upper factor.

2471	2472	2473	2474	2475	2476	2477	2478	2479	2480
$\frac{1}{2471}$	$\frac{1}{2472}$	$\frac{1}{2473}$	$\frac{1}{2474}$	$\frac{1}{2475}$	$\frac{1}{2476}$	$\frac{1}{2477}$	$\frac{1}{2478}$	$\frac{1}{2479}$	$\frac{1}{2480}$

Multiplication, Simple. That in which the factors are terms of the prime series.

Multiplication of Decimals. That in which the factors may consist of positive or negative ranks, or both; and the product shall have as many negative ranks as both the factors.

Multiplication of Mercantile Numbers. That in which products of lower grades are reduced to higher grades by division.

Multiplication of Vulgar Fractions. That in which the product of the numerators is the numerator, and the product of the denominators is the denominator.

N.

Negative Rank. Not a rank, the rank of a fraction, the position of which is in the place occupied by the unit, of which it is a part.

Negative Sign. The dash —, as in subtraction.

Nine. The highest number represented by a single figure, 9; it fills a rank in the decimal scale.

Nominal Moneys. Such as are represented by no coins; as, the pound, livre, and marc banco.

Notation. Any system of marking, the method of forming high numbers by combining the digits.

Nought, or *Cipher* (0). The mark used to fill vacant ranks in the decimal scale.

Number. Any term of the prime series.

Number of Terms. The number which limits a series, upon which limitation the properties of the series partially depend.

Numeration. The reading of numbers.

Numerator. The upper term of a fraction, that which answers to the question, "how many parts?"

O.

Oblique Cross, ×. See Cross.

Odd Numbers. Every second term of the prime series, beginning at 1; as 1, 3, 5, 7, &c.

One. The first term of the prime series, the general name of every single thing, a unit.

Operative Signs. The marks +, ×, —, ÷, √, &c.

Ordinal Numbers. The terms first or prime, second, third, fourth, &c.

P.

Parts. The divisions of a number.

Partial. A term in Alligation, applied to the case in which one of the simples, the rates of the other simples, and also the mean rate, are given to find the quantities of the other simples.

Par of Exchange. The intrinsic equality of different coins and other moneys.

Perfect Numbers. Such as are equal to the sum of all their single parts, thus 6 is equal to its $\frac{1}{2}$, $\frac{1}{3}$, and $\frac{1}{6}$ part; and 28 is equal to

2481	2482	2483	2484	2485	2486	2487	2488	2489	2490
$\frac{1}{2481}$	$\frac{1}{2482}$	$\frac{1}{2483}$	$\frac{1}{2484}$	$\frac{1}{2485}$	$\frac{1}{2486}$	$\frac{1}{2487}$	$\frac{1}{2488}$	$\frac{1}{2489}$	$\frac{1}{2490}$

its $\frac{1}{2}$, $\frac{1}{4}$, $\frac{1}{7}$, $\frac{1}{14}$, and $\frac{1}{28}$ part; and these are all the single parts into which 6 or 28 can be divided.

Per Annum. By the year; per centum, by the 100; per unum, by the single one, or singly.

Period. Six ranks in the decimal scale, counted from the units' point to left and right; the prosodial period, or full point, is used to mark these periods, but especially the units' place.

Permutations. Entire changes in the position of things, with regard to each other.

Perpetuities. Perpetual annuities.

Plus. More, the reading of the erect cross +.

Plural Parts. A fraction, the numerator of which is 2 or more.

Point. See units' point; the first point of the prime series occupies a middle poisition in the decimal scale.

Position. In the prime series, the same as number; because every number has its own position, and but one position in that series.

Positive Rank. The units' rank, or any rank on the left of units, in the decimal scale.

Powers. The successive products of a number multiplied by itself, the number itself being the root or first power, its square the 2d power, its cube the 3d power, &c.

Practice. The rule which computes by parts and complements of parts of the unit of which the price is given.

Premium. A per centum allowance for insurance, brokerage, commission, &c.

Prime. First—prime numbers, multiples of the first term only of the prime series; not composite.

Primordial Series. The series of multiples and the series of single parts of a unit; the series first in order, original.

Principal. The sum put out at interest.

Product. The required term in multiplication.

Progressions. The steps of a series formed by addition, multiplication and their contraries.

Proper Fraction. That in which the numerator is less than the denominator.

Proportions (Analogies). Measured distances upon the line of numbers. Page 93.

Proportionals. The terms of a proportion, viz., two antecedents and two consequents.

Q.

Quantity. A magnitude measured and numbered.

Quotient. The required term in division.

Note. In simple or mercantile numbers, the quotient is a single part of the dividend: in decimal or vulgar fractions, the quotient will be less or greater than the dividend, according as the divisor is greater or less than a unit.

R.

Ranks. Positions in the decimal scale; the places of units, tens,

2491	2492	2493	2494	2495	2496	2497	2498	2499	2500
$\frac{1}{2491}$	$\frac{1}{2492}$	$\frac{1}{2493}$	$\frac{1}{2494}$	$\frac{1}{2495}$	$\frac{1}{2496}$	$\frac{1}{2497}$	$\frac{1}{2498}$	$\frac{1}{2499}$	$\frac{1}{2500}$

100's, &c., on the left, and of 10ths, 100ths, 1000ths, &c., on the right of the units' point.

Rates per Cent. Allowance made upon every 100; as in the case of tares, interest, insurance, &c.

Rates per Unum. Any price, interest, or other allowance made upon things taken singly.

Note. This is the usual basis of calculation, for even when the price is rated by the dozen, groce, &c., it means 1 dozen, 1 groce, &c. The rate per *unum* is the quotient of the rate per *cent.*, divided by 100: and this division is specially taught at page 45.

Ratio. The quotient of a consequent divided by its antecedent.

Rational Root. The root of an exact power.

Real Moneys. Actual Coins, as Eagles, Dollars, &c.

Reduction. The change of measures of all kinds, from one grade of units to another of the same kind.

Remainder. The required term in subtraction, a part of the dividend less than the divisor.

Repetend. The repeating figure in decimals which never terminates, because, however often multiplied by 10, the numerator never can become a multiple of the denominator; as in the case of $\frac{1}{3}$, which equals .333, &c., and $\frac{8}{9}$, which equals .888, &c., without end.

Reversion. The right of an heir to an annuity yet in the possession of another.

Right Angle. The angle of a square. See McCurdy's Euclid.

Root. The first power of a number—the number of terms in a series beginning at 1, is the root or basis of the highest term of that series, whatever that term may be: for thus,

Roots,	1,	2,	3,	4,	5,	6,	&c.
Triangles,	1,	3,	6,	10,	15,	21,	&c.
△ Pyramids,	1,	4,	10,	20,	35,	56,	&c.
Squares,	1,	4,	9,	16,	25,	36,	&c.
□ Pyramids,	1,	5,	14,	30,	55,	91,	&c.
Cubes,	1,	8,	27,	64,	125,	216,	&c.

In which the line of roots shows the number of terms.

S.

Series, Simple. A succession of terms, having a common increase or decrease.

Series, Variable. A succession of terms formed from the terms of a simple series.

Sight. The time of presenting a bill of exchange for payment.

Single Position. A position in the line of numbers, a number, a term of the prime series.

Note. In the rules called single and double positions, *to assume a number* and *to take a position in the line of the prime series* are terms of the same import.

Single Rule of Three. The rule of proportions applied to the common calculations of trade.

Subtrahend. To be subtracted, the less given term in subtraction.

Subtraction. The rule by which is found the excess of one quantity above another.

2501	2502	2503	2504	2505	2506	2507	2508	2509	2510
$\frac{1}{2501}$	$\frac{1}{2502}$	$\frac{1}{2503}$	$\frac{1}{2504}$	$\frac{1}{2505}$	$\frac{1}{2506}$	$\frac{1}{2507}$	$\frac{1}{2508}$	$\frac{1}{2509}$	$\frac{1}{2510}$

Subtraction, Simple. That in which the minuend and subtrahend are terms of the prime series.

Subtraction of Decimals. That in which the two given terms consist of negative ranks alone, or negative and positive ranks in the decimal scale.

Subtraction of Mercantile Numbers. That in which the given terms consist of several grades of units of the same kind, and where a unit of the greater is reduced to make subtraction possible in the less.

Subtraction of Vulgar Fractions. That in which the fraction, or fractions of the minuend, and those of the subtrahend must be reduced to a common denominator before the difference of their numerators can be shown.

Sum. The required term in addition, the amount.

Square. The 2d power of a number, the unit of measure in all superficies. See McCurdy's Euclid.

Synthesis (Composition). The method of reasoning which begins with the simplest elements and proceeds to the more complex combinations.

Surd Root. The root of an affected power, which cannot be exactly ascertained.

T.

Tare. The allowance made for the weight of the barrel, box, cloth, or mat in which goods are packed.

Tens. The 2d positive rank in the decimal scale.

Tenths. The first negative rank in the decimal scale.

Ternary. In combinations, three by three.

Thousands. The 4th positive rank in the decimal scale.

Thousandths. The 3d negative rank in the decimal scale.

Time. That part of duration which is measured by the motion of the celestial bodies, and also by several devices of art.

Troy Weights. The measures of gravitation applied to fine metals, jewels, &c., of which the divisions are the lb., oz., dwt., and gr.

Terms. Words; *termini*, limits: so called, because words define, or limit the range of the idea or image.

The class will please to answer the following questions.

What are the terms used to express the periods of numbers?—the half periods?—the ranks? See pages 12 and 36.

In what rule are the terms *sum*, *aggregate*, and *amount* used?

In what rule are the terms *minuend*, *subtrahend*, and *remainder* used? Also *multiplicand*, *multiplier*, *product*, *factors*, *rectangle? Dividend*, *divisor*, *quotient*, *dividual? Antecedent*, *consequent*, *couplet*, *ratio? Single part*, *complement? Numerator*, *Denominator? Root*, *power*, *index*, *exponent? Principal*, *rate*, *time*, *amount? Extremes*, *means*, *number of terms*, *common difference*, *ratio*, *difference of extremes*, *sum of the series?*

After the foregoing questions are answered, let the meaning of

2511	2512	2513	2514	2515	2516	2517	2518	2519	2520
$\frac{1}{2511}$	$\frac{1}{2512}$	$\frac{1}{2513}$	$\frac{1}{2514}$	$\frac{1}{2515}$	$\frac{1}{2516}$	$\frac{1}{2517}$	$\frac{1}{2518}$	$\frac{1}{2519}$	$\frac{1}{2520}$

each term be given; and also of such other terms as may be supposed to involve the particular illustration of the successive rules; and distinguish the given and required terms in all cases.

U.

Unit. One, the first term of the prime series.
Units. The first positive rank in the decimal scale.
Units' Point. The prosodial period (.), used to mark the first positive rank, and to distinguish the positive ranks from the negative.
Unum. One, *per unum*, singly; the interest of $1.
Usance. "At usance," that is, at the end of a certain number of days, made customary in some European cities, "pay this bill of exchange."

V.

Variable Series. Series formed from the terms of simple series, by consecutive additions.
Vulgar. Common, a term applied only to the general mass of fractions, to distinguish them from those which are decimal.

W.

Weight. Gravity; *Weights*, measures of gravitation, of which three classes are used in commerce; viz., Troy, Avoirdupois, and Apothecaries' Weights.

REMARK.

The trait which most distinguishes this work from all its precursors in the same department of science is the double line of numbers and single fractions which run along the heads of its pages. These lines are indispensable to a perfect system of arithmetic: they afford a mechanical demonstration of every operation; they place before the student a natural image of his thoughts respecting numbers, and a model for the formation of series; and though the allusions to these series are but partial, if the model be followed out in practice, it cannot fail to make intelligent and proficient arithmeticians in the shortest time, and with a comparatively small share of the teacher's assistance.

This is the illustration which seems to have been the aim of the inductive systems; and their falling short of it has left their plans unfinished, and the systems of that class imperfect in the feature in which their chief excellence consists. The primordial series furnish more practice than could be given in several volumes, according to the plan of single questions, and the reason also, in measure and magnitude, for every question that can be proposed. All future attempts to improve arithmetics must converge in the direction of "The New American Order;" and such improvements, if any, must be confined to the detail of particular rules.

CONTENTS.

Proportions used in Simple Interest.

Case 1. $100 : C \times t$, or $1 : u \times t :: p :$ whole interest.

Case 2. $t \times p : a - p :: 1 : u$, or $100 : C$.

Case 3. $p \times u : a - p :: 1 : t$ $a - p =$ wh. int.

Case 4. $1 + (u \times t) : a :: 1 : p$; and
$100 + (C \times t) : C \times t :: a :$ discount.

Note. In every proportion the product of the means divided by one of the extremes quotes the other extreme. The first three terms are the data.

Formulæ used in Arithmetical Progression.

Case 1. $(a+g) \times \frac{1}{2}n = s$; or $\frac{1}{2}(a+g) \times n = s$.

Case 2. $(g-a) \div (n-1) = d$.

Case 3. $a + d \times (n-1) = g$; or $g - d \times (n-1) = a$.

Formulæ used in Geometrical Progression.

Case 1. $(r^{n-1} \times a = g$, and $r^{n-1} \times g = a$.

Case 2. $g + \overline{(g-a) \div (r-1)} = s$.

$a + (g-a) \times \overline{\overline{1 \div (1-r)}} = s$.

$(r^n - a) \div (r-1) = s$.

Formulæ used in Compound Interest.

Case 1. $r^n \times p = a$; and $a - p =$ comp. interest.

Case 2. $a \div r^n = p$, or present worth.

Case 3. $\sqrt[n]{(a \div p)} = r$; because $r^n = a \div p$.

Case 4. $a \div p = r^n$.

Note. Read $\sqrt[n]{(a \div p)}$, the nth root of the quotient of a divided by p; and r^n, r in the power n; or r in the nth power. The letters represent terms which are fully explained in the proper places: it is easy to become familiar with the formulæ, and the advantage to the student will be great.